COMETS

Readings from
SCIENTIFIC
AMERICAN

COMETS

John C. Brandt

NASA-Goddard Space Flight Center
Greenbelt, Maryland
and
University of Maryland
College Park, Maryland

W. H. Freeman and Company
San Francisco

Some of the SCIENTIFIC AMERICAN articles in *Comets* are available as separate Offprints. For a complete list of articles now available as Offprints, write to W. H. Freeman and Company, 660 Market Street, San Francisco, California 94104.

Library of Congress Cataloging in Publication Data

Comets: readings from Scientific American.

Includes bibliographies and index.
1. Comets. I. Brandt, John C. II. Scientific American, Inc.
QB721.C65   523.6   81–4562
ISBN 0–7167–1319–5 (case ed.)   AACR2
ISBN 0–7167–1320–9 (pbk.)

Printed in the United States of America

9 8 7 6 5 4 3 2 1

CONTENTS

Note on cross-references: References to articles included in this book are noted by the title of the article and the page on which it begins; references to articles that are available as Offprints, but are not included here, are noted by the article's title and Offprint number; references to articles published by SCIENTIFIC AMERICAN, but which are not available as Offprints, are noted by the title of the article and the month and year of its publication.

PREFACE

A bright comet suspended in the sky at dawn or dusk is an impressive sight, exciting to scientists and nonscientists alike. The visual spectacle is created by the light scattered from the comet's tail, whose image can stretch from horizon to horizon. The physical tails are normally a few tenths of the distance from the earth to the sun, but in some instances the length has exceeded the earth–sun distance. Studies of the tail show that it is extremely tenuous, comparable to a good laboratory vacuum, and that its total mass is very small indeed.

As our understanding of comets grows, so does our amazement at the diverse physical processes responsible for producing the spectacle of a comet in the sky. All cometary phenomena originate in a central body or nucleus, believed to be only a kilometer or so in diameter and composed of dirty ice. The complex interactions of this ball of dirty ice with the sun's light and expanding solar corona (the solar wind) produce the low density gas and dust tails which scatter the light we see in the sky as a gigantic comet.

The science of comets touches many different fields of study. First, we learn much about the solar wind and about the nature of comets by studying the appearance of comets and their tails. Second, we can learn from the debris of comets that falls on the earth. The smaller pieces, which burn up in the atmosphere, produce most meteors, including those in meteor showers. Occasionally, larger pieces may have hit the earth and produced great destruction. The most famous instance of this is the Tunguska event in Siberia in 1908, which blew down trees within a radius of 30 kilometers. Third, comets can shed light on the early history of the solar system, because they are believed to have condensed from the primitive solar nebula at approximately the same time that the solar system was formed. Because comets spend most of the time in the deepfreeze of space, many scientists believe that their material has been little altered since their formation.

We are in the decade of the once-in-a-lifetime return of the most famous comet of all, Halley's Comet. This comet was the first to have its return predicted. Halley's Comet returns to the inner solar system at intervals averaging 76 years, and its return has been recorded at nearly every apparition since 239 BC. The next apparition will be during 1985–1986, and the prime observing times from earth will be in November of 1985 and April of 1986. Clearly, interest in comets will be at a peak during this period.

The upcoming apparition of Halley's Comet has already sparked great enthusiasm for a space mission to this comet. A spacecraft launched to intercept Halley's Comet could directly measure cometary properties and provide the first images of the surface of the central body of a comet. It is easy to understand the tremendous excitement generated by a mission to a comet and the great advances in cometary knowledge that would result. As of this writing, several missions to Halley's Comet by several nations seem certain.

All aspects of cometary studies have been discussed in articles and other items in *Scientific American*. We begin this reader with an article touching on the general interest in comets as revealed in art and history. One of the fascinating aspects of comets is the public reaction to them. This reaction is revealed in part by the coverage of Halley's Comet at its last apparition in 1910—coverage which included descriptions of the recovery of the comet and of Halley's original work. This comet, and the impression it made on New Yorkers, was the subject of the *Scientific American* cover for April 16, 1910. As to the science of comets, its modern era is generally agreed to have begun in the early 1950s. The articles on comets in *Scientific American* date from July of 1951, and they show the development of cometary astronomy up to the present. Besides these major articles, items from "Science and the Citizen" show the continuing coverage of comets.

An introduction to cometary astronomy precedes this collection of *Scientific American* articles. This introduction will help the reader to appreciate the development of cometary astronomy, its present state, and its prospects. Drawings and photographs illustrate many key concepts throughout this reader. As we approach a period of major activity in the field of cometary research and public interest in comets, this material will provide a useful introduction to the subject.

John C. Brandt
January 1981

COMETS

THE ASTRONOMY OF COMETS

Part of the fascination comets hold for astronomers and the public is their transient behavior. A bright comet with a long glowing tail may be a spectacular sight to the naked eye for a period of several weeks following its discovery, but then it slowly fades out as it recedes toward the cold outer reaches of the solar system. The comet will return again in 76 years if its name is Halley's Comet, but not for another 1,000,000 years or so if it is Comet Kohoutek.

Comets are a bewilderingly disparate group of objects, but despite their variety, astronomers are in agreement on one general point: when carefully studied, comets may provide answers to some profound astrophysical questions, such as the initial composition and formation mechanism of the solar system. Thus, while comets are deserving of study simply "because they are there," the study of comets is related to many other disciplines within, and even outside of, astronomy.

The first observations of comets are, of course, the discoveries. Bright comets are often discovered by amateur astronomers using binoculars or wide-field telescopes. Fainter comets are more often discovered by professional astronomers, usually on wide-field photographic plates taken for some other purpose. The discoveries are communicated to the Bureau for Astronomical Telegrams, Smithsonian Astrophysical Observatory, Cambridge, Massachusetts, and are then announced by International Astronomical Union telegrams.

Normally, comets are named after their discoverers, and, at present, up to three independent codiscoverers are permitted. Comet Kobayashi–Berger–Milon (Figure 9) is an example. Such a rule is easily understood when one realizes that some bright comets have been discovered almost simultaneously by dozens of individuals. Occasionally, a comet has been named after the person who computed its orbit. Examples are Halley's Comet and Encke's Comet.

Comets are designated in two ways. The first comet discovered in 1981 would be 1981a, the second 1981b, and so on for subsequent discoveries. After the orbits have been calculated, comets are assigned a Roman numeral in order of perihelion passage (point of closest approach to the sun). For example, the third comet to pass perihelion in 1981 would be 1981 III. Halley's Comet, at its last appearance, was first Comet 1909c and then Comet 1910 II. Halley's Comet is also designated P/Halley, the P indicating a periodic comet.

Astronomers use many techniques to observe comets. These include visual observations with the unaided eye, binoculars, and telescopes; photographs in visual and ultraviolet wavelengths; photometry (accurate brightness measurements) in visual and infrared wavelengths; spectral scans in many wavelength regions; observations by astronauts orbiting the earth; and observations in extreme ultraviolet wavelengths obtained from rockets and orbiting spacecraft above the earth's atmosphere.

A prerequisite for most cometary work is an orbit determination, for which six parameters are necessary. These parameters specify the orientation of the orbital plane in space, the orientation of the orbit in this plane, the size and shape of the orbit ellipse, and the time of perihelion passage. In principle, three observations of position on the celestial sphere are sufficient to compute an orbit, but in practice, definitive orbits are derived only from many observations.

Accurate orbits have been determined for over 600 comets. Like the orbits of planets, comet orbits are approximately ellipses, but they are much more eccentric (elongated) and have a much greater range of inclinations to the plane of the earth's orbit. Over 500 of these comets are classified as long-period, that is, having orbital periods greater than 200 years. The orbital planes of the long-period comets have approximately random inclinations. This means that there are as many comets in this group with direct or prograde orbits (revolving around the sun in the same direction as the planets) as with retrograde orbits (revolving in the opposite direction from planetary motion). Careful examination of the original (inbound) orbits of the long-period comets shows none that have come from outside the solar system. This situation implies that comets are members of the solar system; no interstellar comets (if they exist) have yet been observed.

The more than 100 short-period comets are mostly in direct orbits with small inclinations. The distribution of periods shows a peak between 7 and 8 years, and the majority have an aphelion distance (point of greatest distance from the sun) near the orbit of Jupiter. This situation implies that the short periods of these comets result from gravitational interactions with the planet Jupiter.

Two other cometary orbital properties are of interest. First, the orbits are not exactly ellipses, with the result that a comet can arrive at perihelion either early or late compared to the prediction based on the previous orbit. The effect, first discovered for Comet Encke, is the result of the so-called nongravitational force. Second, the orbits of many meteor streams are known to be quite similar to the orbits of specific comets. Examples are the η Aquarids, which appear in May, and the Orionids, which appear in October. Both of these showers are associated with Halley's Comet. Thus, the material responsible for the meteor showers is undoubtedly debris from comets.

Spectroscopic observations are the key to understanding the composition of comets. The atoms, molecules, ions, and classes of substances that have been observed are listed in Table 1. Some fairly complex molecules, such as hydrogen cyanide (HCN) and methyl cyanide (CH_3CN), have been found.

TABLE 1.   Species Observed in Comets[a]

Coma (Neutrals)	Tail (Ions)
H, OH, O, S, (H_2O)[b]	CO^+, $CO_2{}^+$, OH^+
C, C_2, C_3, CH, CN, CO, CS	CH^+, $N_2{}^+$, Ca^+, C^+, CN^+
HCN, CH_3CN, NH, NH_2, Na	
Fe,[c] K, Ca, V, Cr, Mn, Co, Ni, Cu	

[a]Silicates are observed in the coma and dust tail.
[b]H_2O has been reported, but the result is controversial.
[c]The heavy metals (such as Fe) are observed when a comet is close to the sun.

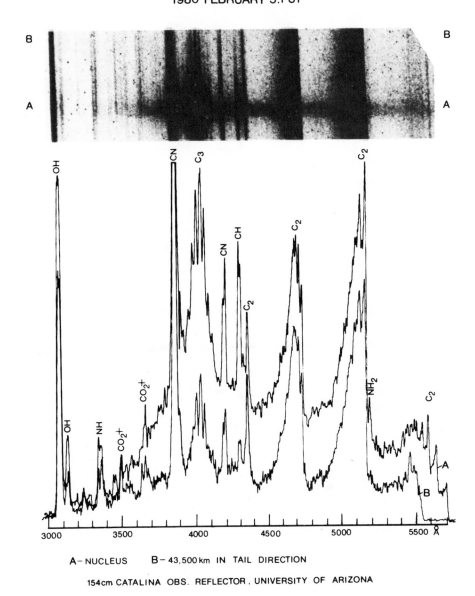

SPECTRUM OF COMET BRADFIELD (1979l)
3000-5500Å

1980 FEBRUARY 5.1 UT

A – NUCLEUS   B – 43,500 km IN TAIL DIRECTION

154cm CATALINA OBS. REFLECTOR, UNIVERSITY OF ARIZONA

Figure 1. Visual wavelength spectrum of Comet Bradfield (1979l) on February 5, 1980, showing many important cometary emissions. (S. M. Larson, University of Arizona.)

Note that water, which (for observational and theoretical reasons) is considered to be the principal constituent of the nucleus, is rarely (if ever) observed directly. However, the table lists H (hydrogen), OH (hydroxyl radical) O (oxygen), H_2O^+ (ionized water molecule), and OH^+ (ionized hydroxyl radical)—all substances having water as a possible parent. Examples of cometary spectra taken from the earth's surface, from sounding rockets, and from orbiting spacecraft are given in Figures 1, 2, and 3, respectively.

Our current knowledge of the structure of comets, from indirect evidence, points to the existence of a central nuclear body or *nucleus* from which all cometary material, both gas and dust, originates. Unfortunately, no photograph of a nucleus exists. Nuclei are probably irregular (see Figure 8), with radii ranging from a few hundred meters to 10 km. Masses range from 100

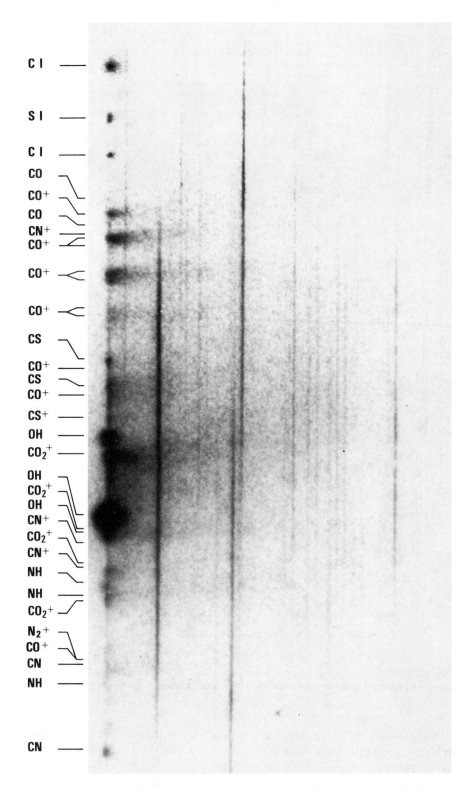

Figure 2. Ultraviolet spectrum of Comet West (1975n) obtained from a rocket-borne instrument flown on March 10, 1976. The coma images are at the left, with the tail images, particularly in CO^+ and CO_2^+, stretching to the right. (A. M. Smith and T. P. Stecher, Laboratory for Astronomy and Solar Physics, NASA-Goddard Space Flight Center.)

million metric tons to 10 trillion metric tons, with roughly equal parts of ices and dust. The average density would be approximately ≤ 2 grams per cubic centimeter. Cometary nuclei sometimes split into two or more pieces; a recent example is Coment West in 1976.

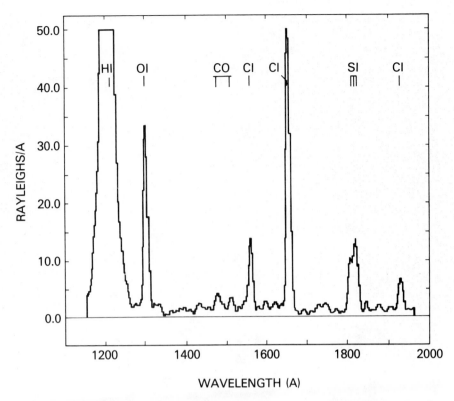

Figure 3. *International Ultraviolet Explorer* observations of the ultravioletspectrum of Comet Bradfield (1971l) in January 1980. The intensity is given in Rayleighs per Ångstrom. Note the large intensity of the resonance line of hydrogen (HI) at 1216 Å. (Adapted from a figure by P. D. Feldman, et al., *Nature* 1980, 286:132.)

The essentially spherical cloud of gas and dust surrounding the nucleus is the *coma*. It can extend as far as 100,000 to 1 million km from the nucleus, and the material flows away from the nucleus at a typical speed of 0.5 km/sec. As the gas flows away from the nucleus, it picks up dust particles through frictional forces, and the dust is dragged outward. For comets at 3 times the earth's distance from the sun (3 astronomical units, AU), the coma is not normally visible and is presumed not to be present. The principal gaseous constituents are the neutral molecules listed in Table 1.

Since 1970, observations from orbiting spacecraft have shown that several comets were surrounded by a giant *hydrogen cloud* extending to 10 million km in diameter, which is larger than the sun. The observations were made in the resonance line of atomic hydrogen at 1216 Ångstroms. A photograph of Comet West's hydrogen cloud is shown in Figure 4, along with a visual photograph to the same scale for comparison. When a bright comet crosses the earth's orbit (a distance of 1 AU from the sun), it emits hydrogen at a rate ranging from ½ metric ton/sec to 1⅓ metric tons/sec. These figures correspond to a total mass loss in the range from 9 metric tons/sec to 24 metric tons/sec.

Photographs of bright comets often show two distinct types of *tails* (Figure 5), the *plasma tail* and the *dust tail*. They can exist separately or together in the same comet. Both point generally away from the sun (the antisolar direction). In a color photograph, the plasma tails appear blue, due to the emission of ionized carbon monoxide (CO^+) bands in the vicinity of 4200 Å (see the photograph on page 55). The dust tails appear yellow because the light from them is reflected sunlight.

The plasma tails are straight and have lengths ranging from 10 million to 100 million km. The plasma in these tails is composed of electrons and molecular ions. The dominant visible ion is CO^+, and the other ions known to be

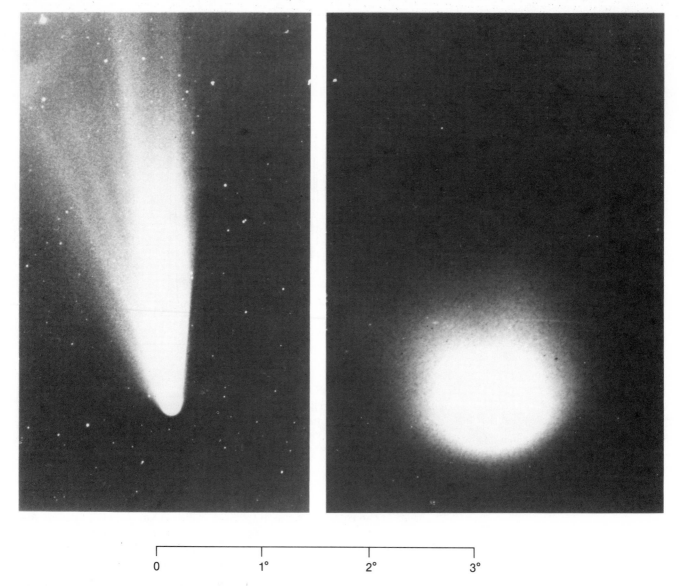

0   1°   2°   3°

Figure 4. Photographs of Comet West (1975n) obtained from a rocket on March 5, 1976 and printed to the same scale. Left, visual light photograph. (P. D. Feldman, Johns Hopkins University.) Right, Lyman-α photograph showing the hydrogen cloud. (C. B. Opal and G. R. Carruthers, Naval Research Laboratory.)

present are listed in Table 1. The production region of the molecular plasma appears to be in the coma near the sunward side of the nucleus. The material in the plasma tails is concentrated into thin bundles or streamers, and this structure is strong evidence for a magnetic field threading the plasma tail. Additional structure is found in the tail in the form of knots and kinks, which appear to move along the tail away from the coma. Plasma tails are usually not observed in comets beyond about 1.5 AU from the sun; an exception is Comet Humason (shown on page 55), which showed a spectacular disturbed plasma tail well beyond the normal distances.

Dust tails are curved and range from 1 million to 10 million km in length. Usually they are relatively homogeneous, but there are exceptions. The available data indicate that the dust particles are approximately 1 micron in diameter and are probably silicate in composition. Occasionally, dust tails are seen which appear to point in the sunward direction, the so-called anti-tails. These are not truly sunward appendages but are the result of projection effects (see the discussion and photographs on pages 45 and 85).

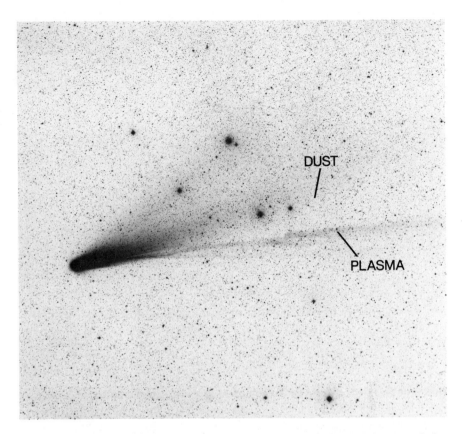

Figure 5. Comet West (1975n) on April 1, 1976. The dust and plasma tails are marked. (Joint Observatory for Cometary Research, operated by NASA-Goddard Space Flight Center and New Mexico Institute of Mining and Technology.)

Figures 6 and 7 show the structure of comets—nucleus, coma, hydrogen cloud, and tails—in condensed form.

Modern cometary theory must explain our current knowledge about the structure of comets. The broad outline of the theory appears to be sound, and we hope to test it directly when spacecraft are sent to a comet. The theory starts with an inactive nucleus, a ball composed of water ice and dust, approaching the sun on its orbit. The surface of the nucleus absorbs energy from

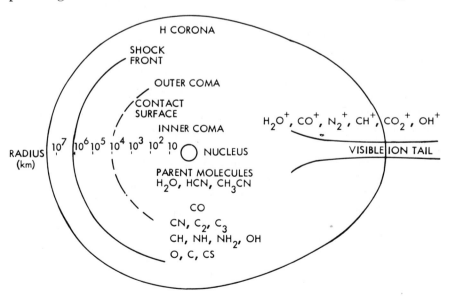

Figure 6. Schematic drawing of the principal gaseous features of a typical comet on a logarithmic scale. (Report of the Comet Science Group, NASA-Jet Propulsion Laboratory, 1979.)

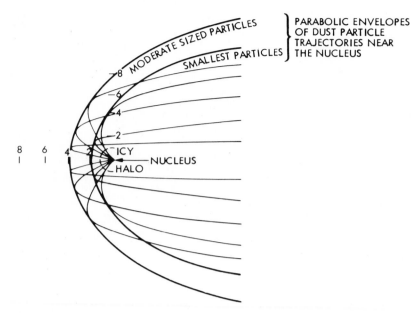

Figure 7. Schematic drawing of the principal particulate features of a typical comet on a linear scale. (Report of the Comet Science Working Group, NASA-Jet Propulsion Laboratory, 1979.)

the sunlight falling on it. When the comet is far from the sun, the energy received simply heats the nucleus. As the comet approaches the sun, however, the temperature of the surface layers increases sufficiently to sublimate the ice from the solid to the gaseous state. When this happens, most of the energy received by the nucleus goes into the sublimation of ices.

If the principal icy constituent of comets is indeed water ice, the theory predicts the onset of cometary activity, as indicated by the appearance of the coma, at approximately 3 AU from the sun. This prediction is in agreement with the observations. The situation is illustrated in Figure 8.

For water ice, the temperature of the sublimating layer would be about 215 K ($-58°C$). Ice fields on the Antarctic Plateau have average tempertures quite close to this value, and these areas could be the site for tests of instruments designed to probe the nuclear surface layers when spacecraft are sent to rendezvous with comets.

The observation that constituents other than water (such as CN, cyanogen) are present when the coma is first observed is explained by the hypothesis that the ice is a special type called a clathrate hydrate. There are cavities in crystalline ice formed by the bonds that hold the water molecules in the crystal. These cavities can be filled by minor constituents up to one-sixth of the crystalline material (water ice) by number of molecules. Thus, when the water ice is sublimated, the minor constituents are released. If there are excess minor constituents in the nucleus, the sublimation process may no longer be strictly controlled by water. As a result, sublimation of the minor constituents could occur well beyond 3 AU on the comet's first approach to the sun. This hypothesis explains why some comets are found to be brighter than expected at great distances from the sun, and the dimming of some comets after their first perihelion passage.

The nongravitational forces have a simple explanation for a sublimating and rotating nucleus. There is a time lag between the maximum solar energy received (local noon on the cometary nucleus) and maximum mass loss through sublimation. Hence, a nonradial reaction force (rocket effect) can be produced, which can speed up or slow down the motion of the comet in its orbit. The physics of nongravitational forces is thoroughly covered in the articles by Fred L. Whipple. In addition, detailed studies of cometary orbits show that

(a)

(b)

Figure 8. Artist's sketches of a comet nucleus. (a) Close to the sun, the nucleus is heated by sunlight, and the sublimation of the ices produces a cloud of gas and dust—the coma—around the nucleus. (b) Far from the sun, the sublimation process ceases and the comet becomes inactive.

they are consistent with the theory of water ice as the substance producing the nongravitational forces.

Once sublimation occurs, the other structural features are produced as follows. The gases, mostly neutral molecules, flow away from the nucleus and drag some of the dust particles with them, forming the coma. Near the nucleus, densities are high enough that chemical reactions can occur. Thus, the molecules observed spectroscopically well away from the nucleus (see Table 1) may not be the same as the gases sublimated from the nucleus. Some of the (presumed) water vapor is dissociated into constituent parts to form the huge hydrogen cloud. To satisfy the details of the observations, dissociation of the hydroxyl radical (OH) may be required as an intermediate step.

The tails are formed in two different ways. The dust tails are relatively simple and consist of dust particles carried away from the nucleus by the flow of coma gases and then blown in the direction away from the sun by solar radiation pressure. The larger dust particles liberated from the nucleus are relatively unaffected by radiation pressure and go into orbit around the sun. These particles reflect sunlight to produce the zodiacal light.

The plasma tails are formed as a result of a fairly complex interaction between the comet and the solar wind (the expanding solar corona). Near the earth, the solar wind consists of a proton–electron gas flowing away from the sun at an average rate of 400 km/sec; this flow carries a magnetic field with it. The interaction is triggered by the presence of ionized molecules in the comet's coma, a situation that usually exists when a comet is within 1.5 AU of the sun. The ionized cometary molecules are trapped and concentrated on the magnetic field lines carried by the solar wind and cause these field lines to slow down in the vicinity of the comet while proceeding at the full solar-wind speed away from the comet. This results in the field lines with the trapped plasma wraping around the nucleus "like a folding umbrella" and forming the plasma tail. The folding can be photographed because the emissions from the trapped ions (such as CO^+) serve as tracers of the field lines. An example of this phenomenon is shown in Figure 9. Although the plasma is indeed swept in the antisolar direction by the solar wind, the plasma tail is a part of the comet usually attached to the near-nuclear region by the magnetic field captured from the solar wind. The exceptions occur when the polarity of the solar-wind magnetic field changes (called a *sector boundary*). This severs the magnetic connection to the near-nuclear region and causes the old plasma tail to detach while the new tail is forming. A dramatic example is shown in Figure 10.

The material used to form the coma, hydrogen cloud, and tails is forever lost to a comet. Roughly 1% of a comet's mass is lost during each perihelion passage. For a comet 1 km in radius, a layer approximately 3 meters thick would be stripped off during each passage. Thus, when the process of sublimation has gone on for a long time, as is the case for the short-period comets, the ices will be exhausted, and a "dead" comet consisting of dust particles and rocks will result. When dispersed along the comet's orbit by gravitational perturbations, the lighter remnants become meteors when they burn up in the earth's atmosphere. The heavier remnants can reach the ground to produce more dramatic results (see page 86).

Figure 11 presents a detailed summary of the physical processes thought to be important in comets.

Where do comets come from? How were they formed? The answer to the first question is probably known. For the long-period comets, the available evidence (from statistics of comet orbits) indicates storage in an approximately spherical cloud around the sun with a radius ranging from 10,000 to 100,000 AU. Gravitational effects from passing stars influence the cloud, limiting its maximum size and randomizing the orbits observed. More importantly, these gravitational effects continually send comets from the cloud into the inner solar system, where they display their coma and tails. Short-period comets most likely are long-period comets that had a gravitational interaction with Jupiter. Recent work indicates that capture into a short-period orbit is the cumulative result of many interactions and usually does not occur on one orbit.

Theories about the formation of comets are much more speculative. The most widely held view is that they condensed from the solar nebula at approximately the same time the sun and planets were formed. Although the details of formation are sketchy, comets are probably a by-product of the solar system's origin, and they may be remnants of the formation process.

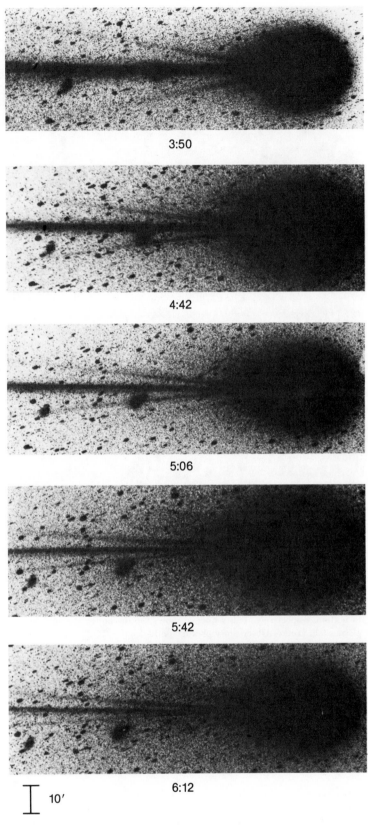

3:50

4:42

5:06

5:42

10′   6:12

Figure 9. Sequence of photographs of Comet Kobayashi–Berger–Milon, July 31, 1975, showing the capture of magnetic field lines from the solar wind. The dominant pair of tail streamers visible on either side of the tail axis lengthen and turn toward the axis in this sequence. (Joint Observatory for Cometary Research, operated by NASA-Goddard Space Flight Center and New Mexico Institute of Mining and Technology.)

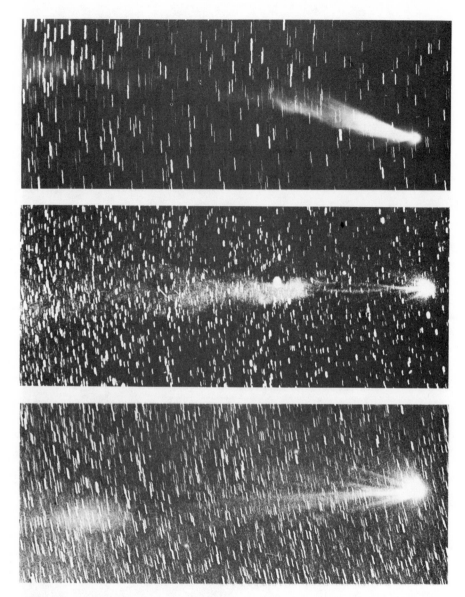

Figure 10. Comet Morehouse (1908 III) on September 30, October 1, and October 2, top to bottom, respectively. This sequence shows the disconnection and drifting away of the plasma tail. (Yerkes Observatory photograph.)

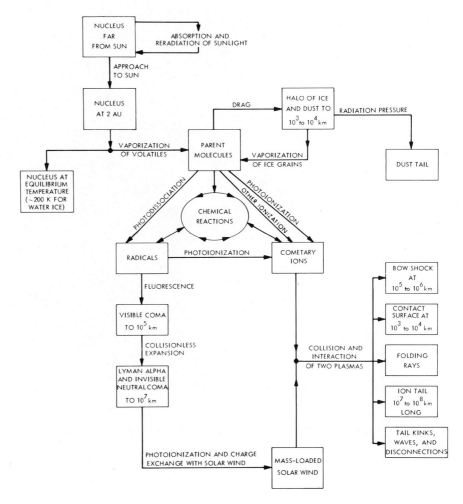

Figure 11. Flow diagram showing features and processes involved in the interaction of a comet with sunlight and the solar wind. (Report of the Comet Science Working Group, NASA-Jet Propulsion Laboratory, 1979.)

INTRODUCTION

Comets are inspiring phenomena to artists, the public, and astronomers alike, and the materials from *Scientific American* collected in this volume clearly show the impact of comets on all three groups.

The strong impression that comets have made on artists through the centuries is visibly demonstrated in the article by Roberta J. M. Olson, "Giotto's Portrait of Halley's Comet." This article also contains a list of all known past perihelion dates of Halley's Comet, as well as a representation of the orbit.

The 1910 appearance of Halley's Comet generated a tremendous amount of public interest and produced some scientific study, which gave an inkling of the diverse physical processes at work in comets. The media response to Halley's Comet in 1910 can be gauged by samples of *New York Times* headlines collected in Figure 1. More remarkable perhaps was the broad *Scientific American* coverage, which ranged from superstition, through the history of cometary studies, to the latest scientific knowledge; coverage even included advertisements for *Scientific American* material related to comets and the cover of the April 16, 1910 issue. The coverage is more appreciated when considered in the context of the nature of *Scientific American* in 1910. The journal was then primarily devoted to the technological progress of the day and featured articles on canals, bridges, ships, electric motors, and so forth; sandwiched in between this material were scientific articles, including the ones on comets. The modern reader should note that there was also a *Scientific American Supplement*, which contained articles also included in this collection.

The coverage began with the August 14, 1909 issue, which presented an article entitled "Halley's Comet" (Volume 101, page 110). This article anticipated the recovery of the comet and alerted the reader to upcoming events. The recovery by M. Wolf on September 11, 1909 was announced in "Return of Halley's Comet—An Ephemeris for This Year" (Supplement No. 1769).

"Could the Earth Collide with a Comet?" (Volume 102, page 194) reviewed a nearly universal concern about comets. An old cartoon treating this subject is shown in Figure 2. The probability of a comet impacting the earth's surface is extremely small, but see page 86 for some idea of what such an event would be like. The next article presented "Condensed Facts About Halley's Comet" (Volume 102, page 258).

One of the highlights of the 1910 *Scientific American* coverage of Halley's Comet was the appearance of articles by Henry Norris Russell, one of America's most distinguished astronomers. Three of his articles are included here. The first, entitled "Halley's Comet at its Brightest" (Volume 102, pages 317–318), is a summary of the observations during the early part of the 1910 apparition. The second article, partially reproduced here, is entitled "The Heavens in May" (Volume 102, page 378). This was an exciting month, because

BARNARD PICTURES OF HALLEY'S COMET

Taken at Yerkes Observatory May 4, They Tally with Observation from Times Tower May 5.

VIEWED BY MISS PROCTOR

Negatives Show the Tail Extending 20 Degrees, Equivalent to 24,000,000 Miles in Length.

IN COMET'S TAIL ON WEDNESDAY

European and American Astronomers Agree the Earth Will Not Suffer in the Passage.

TELL THE TIMES ABOUT IT

And of Proposed Observations— Yerkes Observatory to Use Balloons if the Weather's Cloudy.

TAIL 46,000,000 MILES LONG?

Scarfed in a Filmy Bit of It, We'll Whirl On In Our Dance Through Space, Unharmed, and, Most of Us, Unheeding.

SIX HOURS TO-NIGHT IN THE COMET'S TAIL

Few New Yorkers Likely to Know It by Ocular Demonstration, for It May Be Cloudy.

OUR MILLION-MILE JOURNEY

Takes Us Through 48 Trillion Cubic Miles of the Tail, Weighing All Told Half an Ounce!

BALLOON TRIP TO VIEW COMET.

Aeronaut Harmon Invites College Deans to Join Him in Ascension.

MAY SEE COMET TO-DAY.

Harvard Observers Think It May Be Visible in Afternoon.

MAY BE METEORIC SHOWERS.

Prof. Hall Doubts This, Though, but There's No Danger, Anyway.

YERKES OBSERVATORY READY.

Experts and a Battery of Cameras and Telescopes Already Prepared.

CHICAGO IS TERRIFIED.

Women Are Stopping Up Doors and Windows to Keep Out Cyanogen.

Figure 1. Facsimile newspaper headlines from the 1910 apparition of Halley's Comet, *New York Times*, May 10, 16, and 18. (Copyright © 1910 by the New York Times Company. Reprinted by permission.)

not only was the comet a spectacular sight in the sky, but it transited the sun on May 18, and the earth either passed through the tail or just missed it one or two days later. A separate article, "The Transit of Halley's Comet" (Volume 102, page 438), discussed the event. The third article by Russell, partially reproduced here, is "The Heavens in July" (Volume 103, page 12). Halley's Comet had become faint in the sky in July, and public interest had also begun to wane at that time.

The final two articles from this period concern the history of cometary science. "Halley's Cometary Studies—His Own Account of His Investigations on Orbits" (Supplement No. 1782) describes the original work that led to Halley's prediction in 1705 that the comet which now bears his name would return in 1758–1759. Finally, "Edmund Halley, The Man Who Dispelled Com-

Figure 2. A French cartoon warning that a comet would hit the earth on June 13, 1857.

etary Superstitions" (Supplement No. 1770) is a biographical sketch of a man considered by some to be the greatest astronomer of all time (Figure 3).

The fascination with Halley's Comet has probably come through clearly in the material discussed so far. Halley's is the most famous of all comets for several reasons. It was the first to have its return predicted, and the average orbital period of 76 years is comparable to the human life span. Also, the appearance of the comet through the centuries has been associated with many important events in history. In addition, Halley's large size, the full range of activity it displays, and its orbit with perihelion favorably placed about half way between the sun and the earth's orbit tend to ensure naked-eye visibility of the comet at each apparition. Unfortunately, the upcoming passage in 1985–1986 may be an exception.

The next perihelion passage is expected on February 9, 1986, when, along with the scientific interest, massive media and public attention is expected. A recent indication of the popular interest in comets came with the appearance of Comet Kohoutek. During the three-week period in January 1974 when the comet was brightest, the Hayden Planetarium of the American Museum of Natural History in New York received some 1,000 phone calls per day and a total of approximately 20,000 letters. There is no reason to believe that public response will be less enthusiastic in 1985 and 1986.

Halley's work elevated the scientific study of comets to a respectable level and gave it a firm footing. He built on the following developments. Prior to 1577, comets were considered to be phenomena in the earth's atmosphere, a view promulgated by Aristotle in the fourth century BC. The Aristotelian view would dominate the scene for approximately 2,000 years, even though some thinkers (such as Seneca in his *Natural Questions*) would write about comets in surprisingly modern terms. In 1577, however, Tycho Brahe showed conclusively that comets were not atmospheric. He searched for a parallax of the Comet of 1577 in the position among the stars when it was observed high and low in the sky and when it was observed between two different locations on earth. An atmospheric object should have had an easily measured parallax, but Tycho found none. Given his reputation for accuracy, when Tycho cautiously concluded that the comet was at least four times the moon's distance,

Figure 3. Edmund Halley, 1656–1742.

it was accepted that comets were celestial and not atmospheric phenomena.

Subsequent research focused on the orbits of comets. It was found that parabolas accurately represented the orbits of most comets near the sun. The first comet to which Halley applied Newton's work on gravitation was the Comet of 1680, and he derived the elliptical orbit published in the latter's *Principia*. Halley also noted the orbital similarity of comets observed in 1531, 1607, and 1682, and he concluded that the observations referred to a single comet with a period of about 76 years. Thus, he could predict its return in 1758–1759.

In the nineteenth century, more information was obtained relating to cometary orbits. The phenomenon of a steadily changing orbital period—commonplace enough today, but unknown in the early 1800s—was discovered in Comet Encke. This characteristic is attributed to the so-called nongravitational forces. Further work included the discovery of a close connection between the orbits of comets and meteor streams. The physical study of comets also began in the nineteenth century. Bessel investigated the forms of comet tails (stimulated by the appearance of Halley's Comet in 1835), and this work was extended by Bredichin.

Progress in the first half of the twentieth century was sparked by extensive observations of Comet Morehouse in 1908 and Comet Halley in 1910. Despite the accumulation of a great deal of information, however, the foundation on which an understanding of comets could be built would not be available until the second half of the twentieth century.

There is widespread agreement that the basis for modern cometary science began with three papers published in 1950 and 1951, and there is excellent coverage of these events in *Scientific American.*

(1) In 1950, J. Oort proposed the currently accepted answer to the question "Where do comets come from?" He developed the idea that comets are in a spherical cloud a large distance from the sun, a "storehouse." Occasional gravitational perturbations by passing stars send comets into the inner solar system, where we observe them. Naturally, this process leads to the eventual destruction of the comet, because passages near the sun cause an irretrievable loss of material. Oort's theory pervades all the modern (post-1950) *Scientific American* articles presented here.

(2) In 1950, Fred L. Whipple proposed his icy conglomerate model of the cometary nucleus. In this model, the nucleus was not a cloud of particles (the "sand-bank" model), as previously thought, but a single mass of ice and dust particles, commonly called a "dirty snowball." Such a nucleus could supply an adequate amount of gas to explain observed cometary phenomena and last through many apparitions, because only a thin surface layer would be eroded away by sublimation during each passage near the sun. Whipple describes the initial development of these ideas in his 1950 article "Comets." The principal opposing idea, the "sand-bank" model, was presented by R. A. Lyttleton in his book *The Comets and Their Origin.* James R. Newman, in his review of Lyttleton's book, presents Lyttleton's theory of the origin of the sand-bank. While the sand-bank model was accepted for decades and has appeared in many textbooks, it is no longer tenable. Newman's review is also a nice presentation of many other facets of cometary astronomy.

(3) In 1951, Ludwig F. Biermann postulated the existence of a continuous outflow of ionized gas from the sun. He called this gas the solar "corpuscular radiation," which we today call the solar wind. Biermann was led to his postulate by studies of features in plasma tails, particularly their observed accelerations away from the sun. These ideas are discussed in "The Tails of Comets" by Biermann and Rhea Lüst. Near the end of the article, the possible role of the magnetic field carried by the solar wind (as suggested by H. Alfvén) is mentioned. It is now clear that the complicated interaction of the solar wind with comets plays an important role in the formation of plasma tails and many other aspects of cometary physics.

The appearance of Comet Kohoutek in 1973–1974 produced new enthusiasm for the study of comets. The field was reviewed in Fred L. Whipple's 1974 article "The Nature of Comets." The article included a review of basic comet models (sand- or gravel-bank versus the dirty snowball) and a thorough discussion of the nongravitational forces. Water ice is noted to be the probable major volatile constituent of the nucleus, and the possibility of a space probe to a comet is briefly mentioned.

In Fred L. Whipple's 1980 article "The Spin of Comets," we find an extensive discussion of Comet Encke, and especially of its changing period and the forces responsible for the changes. In addition, this article brings us to the bittersweet subject of space missions to comets. The sweet part is the opportunity to establish some definitive vital information that appears to be obtainable in no other way. For example, we have no photographic image of a comet nucleus, and we have no in situ measurements of a comet's magnetic field or density in the coma. In other words, by current standards of space exploration of the solar system, comets are unexplored. Exploration efforts have predictably focused on the return of Halley's Comet in 1985–1986.

The bitter part of this subject is the lack (as of January 1981) of any approved NASA space mission to a comet. The Halley Flyby/Tempel 2 Rendezvous Mission described by Whipple is no longer possible, nor is a mission studied earlier to rendezvous with Halley's Comet itself. Both these missions required the development of a new propulsion system known variously as solar-electric propulsion or "ion drive," but approval for the rush development of this system was not forthcoming. However, an excellent pure flyby mission called the *Halley Intercept* can still be developed and launched to Halley's Comet if approval is granted within the next few months. Besides the capability for in situ measurements, a major feature of the *Halley Intercept* mission is an imaging capability comparable to the recent *Voyager* missions to Jupiter and Saturn. The system envisioned would obtain spectacular views of the entire comet while the spacecraft was in the "cruise phase," and the first photographs of the nucleus at closest approach. It could also provide imaging coverage of the comet at perihelion when it is brightest but when viewing conditions from earth are unfavorable (see below).

In contrast, flyby missions to Halley's Comet have already been approved by Japan and by the European Space Agency. The latter's mission has been named *Giotto* (for the significance of this name, see the article by Roberta J. M. Olson). A flyby mission is also under consideration by the Soviet Union.

Despite the uncertainty of U.S. plans to mount a space mission to Halley's Comet, the 1985–1986 apparition should be a time of great interest and scientific advances. The path of Halley's Comet on the sky in 1985–1986 is shown in Figure 4, and a condensed presentation of observing conditions is shown in Figure 5. The prime Northern Hemisphere observing period is in November of 1985, and the prime Southern Hemisphere observing period will be in April of 1986. NASA has organized a coordinated, world-wide effort called the "International Halley Watch," with the goal of stimulating, encouraging, and coordinating observations during the apparition. Detailed viewing instructions will be issued by the International Halley Watch (NASA/Jet Propulsion Laboratory, Pasadena, California) as the time approaches. Note in Figure 5 the circumstances accompanying maximum brightness (*minimum* apparent total magnitude) at perihelion, February 9, 1986. Observing conditions are terrible because the comet and earth are essentially on opposite sides of the sun, as shown on page 8. The inevitable gap in the ground-based coverage resulting from the unfavorable geometry at the 1985–1986 apparition could be filled

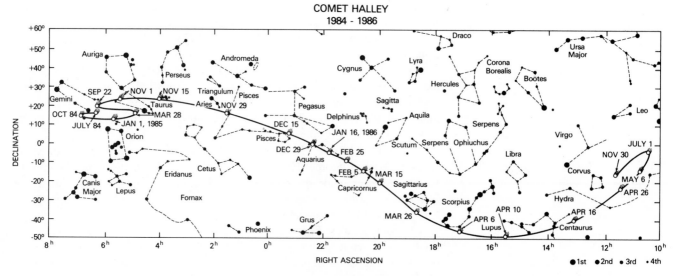

Figure 4. Path of Halley's Comet on the sky from July 1984 to November 1986. (Laboratory for Astronomy and Solar Physics, NASA-Goddard Space Flight Center.)

by a space mission. If successful, the imaging system aboard the *Halley Intercept* could provide coverage of great interest to both the public and the scientific community.

In "The Origin and Evolution of the Solar System" by A. G. W. Cameron, the author presents his theory of solar system formation, including the formation of comets. The physical origin of comets is a very difficult subject, and the reader should be aware (as Cameron himself notes) that widely differing views exist.

In "The Solar Wind" by E. N. Parker, we encounter another use for comets, that is, as natural (and "free") probes of the solar wind. Many scientists believe that Biermann's work in the early 1950s on comet tails included the discovery of the solar wind. Parker's article accurately represents the state of knowledge of the solar wind in 1964, and the reader should bear in mind that much has happened since then. Nevertheless, the use of comets as solar-wind probes continues to be an active area of research.

Comets have been the subject of news items in *Scientific American* since 1950. Several of these are included here, together with a short Introduction.

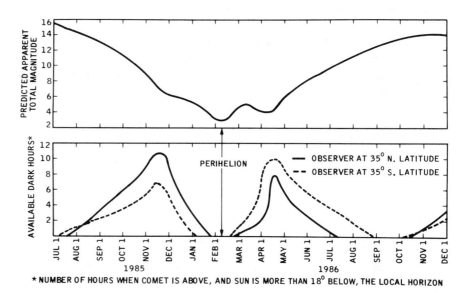

* NUMBER OF HOURS WHEN COMET IS ABOVE, AND SUN IS MORE THAN 18° BELOW, THE LOCAL HORIZON

Figure 5. Observing conditions for Comet Halley. (NASA-Jet Propulsion Laboratory.)

Giotto's Portrait of Halley's Comet

by Roberta J. M. Olson
July 1979

A blazing comet represents the star of Bethlehem in one of Giotto's famous Arena Chapel frescoes in Padua. It is a naturalistic portrait of Halley's Comet as seen during its spectacular apparition of 1301

In the late summer of 1301 a spectacular comet splashed across the dome of the night sky. Its luminous spherical coma surrounded a starlike central incandescence that seemed to spew a diffused flare of radiant particles, like a mane of hair, directly away from the sun. We know now that it was Halley's Comet, making one of its periodic visits to the center of the solar system. And we know how it looked to amazed observers in Italy because its "portrait" was rendered in faithful detail by Giotto di Bondone (1267–1337), the Florentine pioneer of naturalistic painting.

Giotto was in Italy in 1301 (although exactly where is not certain), and he surely saw the comet. Within perhaps a year and no more than four years he executed the remarkable series of frescoes on the walls of the Scrovegni (Arena) Chapel in Padua, in northern Italy. In a scene depicting the Adoration of the Magi he represented the star of Bethlehem, in a remarkable departure from iconographic tradition, not as a stylized, many-pointed little star but as a blazing comet. The coincidence of dates, the naturalistic representation of the comet and its similarity to photographs of the comet made on its most recent appearance, in 1910, constitute strong evidence that Giotto's comet is indeed Halley's Comet.

There are billions of comets (the word is from the Greek *kometes*, "long-haired") orbiting the sun in a thinly distributed cloud that extends up to thousands of times as far as the outermost planets. Each comet nucleus is a small, icy aggregate of frozen gases and interstellar dust. Occasionally one of these "dirty snowballs," its orbit perhaps having been perturbed by a star, enters the inner regions of the solar system, where its gases and dust, activated by solar radiation and the solar wind, form a visible coma, or head and tail [see "The Nature of Comets," by Fred L. Whipple; page 47]. Some of these comets attain a new elliptical orbit that periodically brings them within sight of the earth.

Halley's Comet was the first comet to be recognized as being periodic. In 1705 the English astronomer Edmund Halley noted striking similarities in the orbits he had calculated, based on earlier observations, for major comet appearances in 1682, 1607 and 1531. He postulated that the three apparitions represented return visits, approximately 76 years apart, of a single comet describing an elongated elliptical orbit around the sun. He predicted that it would return in 1758. Later astronomers, allowing for perturbations of the orbit by Jupiter and other planets, refined Halley's calculations to predict that the next perihelion passage (the closest approach to the sun) of Halley's Comet would be in April of 1759. The comet was first sighted by a German peasant on Christmas Day in 1758, and it reached perihelion in March of 1759.

Subsequent calculations have established the date of every perihelion passage back to 239 B.C.; examination of old records (largely those of Chinese observers for passages before the 15th century) reveals observations of a comet at the right time and in the right part of the sky to match most of the calculated returns to perihelion. The mean period of the comet is just under 77 years, but the precise period varies as much as two and a half years on each side of the mean because of perturbations by the planets. Halley's Comet will return to view during 1985, with perihelion predicted for February 9, 1986.

It is estimated that in every century perhaps 20 or 30 comets are visible from the earth. In the 13th century, two centuries before the invention of the telescope, more than 20 unaided-eye sightings of "comets" were recorded. Most of the "comets" mentioned by Western chroniclers are referred to only once, however, and without precision; some of them were certainly meteors or other celestial phenomena. The only reliable comet apparitions are those that have been verified by careful correlation of several observations, including sightings recorded in Chinese annals brought to the West by returning Jesuit missionaries in the 18th century and first translated in 1846. Presumably these were the most spectacular apparitions of their time, since they drew enough attention to give rise to multiple references in Europe and in Asia. A meteor might be recorded as a comet by one person, but a genuine comet would not be likely to confuse several witnesses. A spectacular apparition would astonish all observers.

Of the comets with calculated orbits securely recorded for the 13th century only two have been established as spectacular apparitions. The first was seen over both China and Europe in 1264. The second was definitely seen over China in 1299 (and a treatise attributed to one Peter of Limoges records such a comet over Europe). Giotto was born in about 1267, and so he could not have observed the comet of 1264. The apparition of 1299 probably was not seen over Italy or was unremarkable there. The first securely established spectacular apparition within Giotto's lifetime was therefore that of Halley's Comet in 1301. The next spectacular comet apparition was not until 1337, the year of Giotto's death. Moreover, the apparition of 1301 was an impressive one, commented on by many contemporary historians. The eminent 14th-century Florentine chronicler Giovanni Villani wrote in his *Chroniche Storiche* that a comet appeared in the heavens in September of that year "with great trails of fumes behind" and remained visible until January of 1302. There are various discrepancies in the dating, with some writers reporting that the apparition remained for only six weeks, but all agreed that the comet's tail was impressively long; it is estimated to have subtended an angle of as much as 70 degrees in the sky. Such an apparition accords well with Giotto's representation.

Throughout recorded history the appearance of a comet has been an occasion for fear. For millenniums it was believed that all aspects of life on the earth were ruled by the positions of the stars and planets. Once the locations of the fixed stars and the changing positions of the planets were known the appearance of a "new star" of any kind seemed to violate the order of the heavens and hence to be prescient of disaster (literally a "bad star"). Such appearances were often associated with significant human events, yet before the 16th century comets were only rarely represented in Western art. (One example is a sword-shaped comet shown hovering in the sky on a Roman cameo.) In medieval times comets are sometimes seen in subsidiary positions in large astrological works, such as the 12th-century zodiacal cycle on the west facade of the Cathedral of Piacenza in central Italy. Such representations of comets are generic, however; "portraits" of identifiable comets—explicit representations of historical apparitions—are even rarer.

Three such "portraits" of apparitions of Halley's Comet before Giotto's time have been identified. The earliest one represents the apparition of A.D. 684, although it was not executed until eight centuries after the fact, in the *Liber Chronicorum,* or *Weltchronik,* of Hartmann Schedel, known in English as the *Nuremberg Chronicles* because it was printed in Nuremberg. The book was first issued in 1493 and was illustrated with woodcuts by the German artist Michel Wohlgemuth and his stepson Wilhelm Pleydenwurff. The printer was Anton Koberger, the godfather of Albrecht Dürer (who was later to portray a spec-

THE ADORATION OF THE MAGI is one scene in a fresco cycle executed by the Florentine master Giotto di Bondone. The cycle decorates the interior of the Arena Chapel, which was commissioned by the Paduan businessman Enrico Scrovegni (perhaps to expiate the sin of his father, who was identified by Dante in the *Inferno* as the archusurer). Scrovegni obtained permission to erect the building in 1302, the site was dedicated in 1303 and the frescoes appear to have been begun that year; the *Adoration,* on the second tier of the many-tiered cycle, was probably completed in about 1304. The scene exemplifies Giotto's major innovations, his naturalism and the humanity of his figures, and is notable for its representation of the star of Bethlehem not as a stylized star but as a dynamic comet. Halley's Comet, which returns to center of solar system about every 77 years, made an appearance in 1301. It served as the model for Giotto's comet.

CLOSE-UP OF COMET in the *Adoration* shows how the artist applied tempera and gold pigments to the plastered wall in textured strokes approximating the luminescent appearance of the coma and tail of the actual comet, which he must have observed carefully a few years earlier. He depicted the intensely bright center of the coma, or head, with what appears to be an eight-pointed star, building up layers of pigment over the star to diffuse the image. Some pigment is lost, revealing the red adhesive by which it was applied to the plaster.

tacular comet apparition in his engraving *Melencolia I*). A crude representation of Halley's Comet appears with accompanying text on the page dealing with the year 684. (It is repeated haphazardly throughout the book because the limited repertory of small woodblock designs functioned not only as specific illustrations but also as "finger posts" for the reader searching for his place in the large unpaged volume.)

The earliest contemporaneous, albeit stylized, portrait of Halley's Comet depicts the apparition in the spring of 1066. It is found in a scene of the Bayeux Tapestry, which was commissioned by Queen Matilda, the wife of William the Conqueror, to illustrate her husband's victory at the Battle of Hastings. The tapestry (actually a crewel embroidery on eight narrow strips of coarse linen,

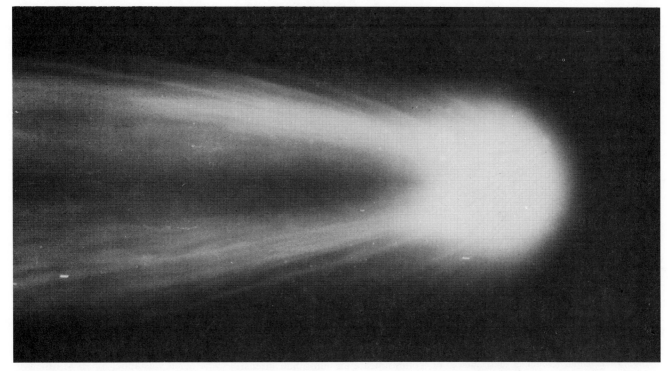

HEAD OF HALLEY'S COMET was photographed with the 60-inch reflecting telescope on Mount Wilson on May 8, 1910, during the comet's most recent return to the vicinity of the earth. Major comets have distinctive, although somewhat variable, configurations, and this photograph is recognizably similar to Giotto's depiction. The comet proper is a small (and hence invisible) nucleus of frozen gases and dust. When the comet comes near enough to the sun, some of the gas is evaporated, carrying dust particles with it; the glow of the comet's coma is from sunlight that is scattered by the dust or reradiated by fluorescing gases. The tail, which always points away from the sun, is a stream of dust impelled by solar radiation pressure and of gas molecules ionized by solar radiation and by electrons in the solar wind and impelled by magnetic fields within the solar wind. Halley's Comet will return to perihelion in 1986 (*see illustration on page 8*).

RETURNS TO PERIHELION OF HALLEY'S COMET	
239 B.C.	MARCH 30
163	OCTOBER 5
86	AUGUST 2
11	OCTOBER 5
A.D. 66	JANUARY 26
141	MARCH 20
218	MAY 17
295	APRIL 20
374	FEBRUARY 16
451	JUNE 24
530	SEPTEMBER 25
607	MARCH 13
684	SEPTEMBER 28
760	MAY 22
837	FEBRUARY 27
912	JULY 9
989	SEPTEMBER 9
1066	MARCH 23
1145	APRIL 22
1222	OCTOBER 1
1301	OCTOBER 23
1378	NOVEMBER 9
1456	JUNE 9
1531	AUGUST 25
1607	OCTOBER 27
1682	SEPTEMBER 15
1759	MARCH 13
1835	NOVEMBER 16
1910	APRIL 20

RETURNS TO PERIHELION, or closest approach to the sun, have been calculated back to 239 B.C. and most of the dates have been correlated with records of observations. The average period of the comet's long elliptical orbit is almost 77 years, but successive periods vary as the orbit is perturbed by the gravitational effect of planets, Jupiter in particular.

Cometa

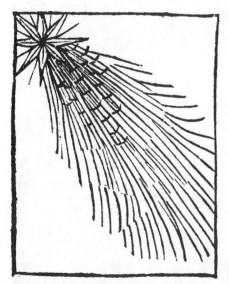

NUREMBERG CHRONICLES, published in 1493, contains this woodblock illustration of a comet on the page recounting the events of A.D. 684, a year in which Halley's Comet appeared. The accompanying text tells of calamitous events brought about by the comet: three months of rain, thunder and lightning, during which people and flocks died, grain withered in the fields and an eclipse of the sun and moon was followed by a plague.

measuring almost 231 feet long by 19½ inches wide) was executed between 1073 and 1083 and now hangs in the town hall of Bayeux in Normandy. Numerous accounts by contemporary chroniclers date the 1066 apparition and remark on its spectacular nature, which is attested to in the tapestry by the amazed expressions on the faces of its beholders and by the embroidered legend: "They are in awe of the star." The comet itself is represented, in the prevailing Romanesque style of the period, as a highly stylized geometric structure, a flat and purely decorative configuration of lines and planes.

The next portrait of Halley's Comet is found in the *Eadwine (Canterbury) Psalter*, a 12th-century English manuscript that is a copy of the ninth-century *Utrecht Psalter*, a collection of the Psalms. At the bottom of the page giving the text of Psalm 5 the comet is seen in bare outline: a stylized rosette inscribed in a circle, with a tail of four wavy rays. The position of the sketch at the bottom of the page and its large scale suggest that it was added, along with its Old English textual description, as a passing notation; it appears to have no connection with the three Latin versions of the Psalm on the page. The style of the manuscript and the dates of its scribe, the monk Eadwine, correspond with the 1145 apparition of Halley's Comet. In spite of its sketchiness this drawing is revolutionary for the 12th century in that it is a contemporary representation of a natural phenomenon.

The most remarkable aspect of Giotto's portrait of Halley's Comet, in sharp contrast to the schematic nature of the earlier representations, is its naturalism. This is fitting, because Giotto's contemporary reputation and his immense significance in the history of painting (he was the first artist to exert an almost universal influence on Western painting) proceed from his startlingly naturalistic innovations. He was born in the small village of Colle di Vespignano, near Florence. Little is known of his early life, but he is thought to have been a pupil of the Florentine painter Cimabue, a progressive master who began to break with the prevailing Italo-Byzantine style. Giotto rebelled more dramatically against the established style and developed a new three-dimensional plasticity that must have seemed even more overwhelming in his day than it does to connoisseurs today. His majestic sculptural figures, calmly positioned to form simple masses of glowing color, communicate an emotional intensity unprecedented in European painting. And it was in the chapel frescoes in Padua that Giotto's revolutionary style achieved not only maturity but also its most lucid statement.

Giotto had worked in Rome at about the turn of the century and is thought to

have been in Padua by 1302, working on a fresco that has since been lost. A wealthy Paduan merchant, Enrico Scrovegni, commissioned him to decorate the interior of a small family chapel being built adjacent to the family palace and near the ruins of a Roman amphitheatre (whence the familiar name Arena Chapel). It has been suggested that Giotto was the architect of the chapel, in which case the conceptual date of the frescoes would be even closer to the apparition of the comet; in any case they appear to have been started in 1303 at the latest. Giotto covered the walls with tiered rows of scenes illustrating the life of Christ, from the events before his birth (the lives of St. Joachim and St. Anne, parents of the Virgin Mary, and the life of the Virgin) to the infancy of Christ, his adult life, his passion, crucifixion and resurrection and the Last Judgment, all within a larger symbolic program. Most of the 38 narrative scenes, including the *Adoration,* measure about 200 by 185 centimeters, with figures about half life size. Because of its position in the second tier the *Adoration* can be dated to 1303 or 1304.

When Giotto came to paint the star of Bethlehem, he rejected the strictures of both astrological symbolism and medieval convention and rendered the comet as he had actually seen it a few years before, theatrically illuminating the Italian night sky. The large, fiery comet dominates the sky of his fresco. The coma pulses with energy; in its center is a hard-edged star shape representing the bright "center of condensation" that is often seen within the more diffuse coma. (It is not the nucleus, which is too small to be visible within the coma.) The striated tail imparts a dynamic impression of the arc traced by the comet's passage across the sky. Spectacular comets do look like this to the unaided eye, and Halley's Comet would have looked like this to Giotto. One conjectures that he might have recorded his observations of 1301 in a drawing to which he referred later (although no drawings made by him are extant). Whether that was in fact the case or whether he simply recalled the apparition vividly, his naturalistic comet harmonizes perfectly with the naturalistic aesthetic that characterizes the *Adoration* and the entire cycle of frescoes.

What could have motivated Giotto to represent the star of Bethlehem as a comet? No one knows, of course, whether the biblical star was a comet, a nova or a conjunction of planets, or whether it was apocryphal. Matthew is the only Evangelist who mentions a star leading the Magi to the Nativity, and he gives no detail: "For we have seen his star in the east, and are come to worship him.... And, lo, the star, which they saw in the east, went before them, till it came and stood over where the young child was. When they saw the star, they rejoiced with exceeding great joy" (Matt. 2:2; 2:9–10). The star is also mentioned in the apocryphal *Protoevangelium of James,* but again without characterization. For centuries, however, the star of Bethlehem was depicted in Nativity and Adoration scenes. Almost always it was shown as a small, stylized and quite imaginary star, often with rays of light shining down on the Infant Jesus, signifying God's blessing on the birth.

Renaissance painters depended on well-established rules of iconography. Style and technique might vary, but the choice of what to put in a painting, even where to place the figures and how to pose them, was largely established by tradition (or innovative interpretations of tradition), which in turn shaped current expectations. Giotto was therefore not likely to have substituted a comet for the conventional star on mere impulse, even in response to a vivid personal observation.

Certainly there was a lengthy literary tradition concerning comets. Long before Giotto such writers as Aristotle, Virgil, Seneca and the Roman poet Lucan, among others, had speculated about comet apparitions. In general comets were regarded with apprehension and interpreted as portents of catastrophe, plagues or the death of kings. Less often they were considered positive signs—of victory, bounty or the birth of kings. The interpretation varied with circumstance. The apparition of Halley's Comet in 1066 could have been taken by the English as a portent either of their King Harold's victory over Harold III of Norway in late September of that year or of his defeat the following month by William of Normandy at Hastings; the Normans, the eventual victors in the three-way conflict, obviously came to regard the comet as being prescient of William's triumph. Comets that were regarded as positive omens were generally believed to have been of divine rather than natural origin—created by God for a particular benevolent purpose. This view was echoed in Giotto's own time by Aegidius of Lessines, in his treatise of 1264 *On the Essence, Motion and Signification of Comets.*

At the time of the writing of the Gospels at the end of the first century it was common to associate the appearance of a "new star" with the birth of a king. Much later the writings of the church fathers, which would have been familiar to Giotto's contemporaries, expanded on Matthew's account of the star of Bethlehem. Two innovative Christian thinkers, Origen in the third century and John of Damascus in the seventh and eighth centuries, remarked on the cometlike nature of the star described by the Evangelist.

In his treatise *Against Celsus* Origen wrote: "The star that was seen in the

NATIVITY BY DUCCIO DI BUONINSEGNA is a small panel from the predella, or base, of a very large altarpiece, the *Maestà,* executed between 1308 and 1311 for the Siena Cathedral; the panel (with two side panels, not shown here) is now in the National Gallery of Art in Washington. Duccio's star of Bethlehem, in contrast to the comet in Giotto's *Adoration,* is a small star with faint rays directed toward the Child. Duccio was Giotto's senior, from a rival school.

NATIVITY BY A FOLLOWER OF GIOTTO was painted in about 1316 in the Lower Church of San Francesco in Assisi; the fresco, of which this is a detail, is a pastiche of motifs from scenes in the Arena Chapel. There is a comet, but it is small and stylized, a pale echo of Giotto's, and it has been subordinated to the traditional heavenly rays descending to the Child below.

east we consider to have been a new star . . . such as comets, or those meteors [comets and meteors were not then differentiated] which resemble beams of wood, or beards or wine jars." And he pointed out that "we have read in the *Treatise on Comets* by Chaeremon the Stoic, that on some occasions also, when *good* was to happen comets made their appearance. . . . If then, at the commencement of new dynasties . . . there arises a comet . . . , why should it be a matter of wonder that at the birth of Him who was to introduce a new doctrine to the human race . . . a star should have arisen?" Origen acknowledged that there was no specific early prophecy that a particular comet would "arise in connection with a particular kingdom or a particular time, but with respect to the appearance of a star at the birth of Jesus

BAYEUX TAPESTRY, which commemorates the events of 1066, records the apparition of Halley's Comet in the spring of that year. At the left a crowd of Englishmen point to the stylized comet in the upper border; the legend above them reads: "They are in awe of the star." At right King Harold II of England, on being told of bad omen, imagines ghostly invasion ships (*bottom*) that foreshadow his defeat.

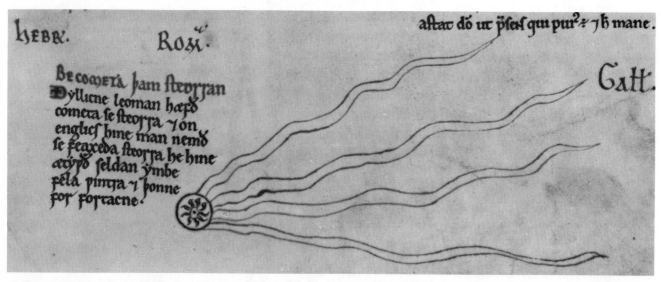

EADWINE PSALTER, an illuminated-manuscript collection of the Psalms, was copied by the monk Eadwine from the earlier *Utrecht Psalter*. Judging from Eadwine's dates, stylistic evidence and this representation of a comet at the bottom of a page, the copy was made in or soon after 1145, when Halley's Comet appeared. The legend refers to the radiance of the comet, or "hairy star," and remarks that comets appear rarely, "and then as a portent." Above comet are three Latin versions of Psalm 5: *Hebraicum, Romanum* and *Gallicanum*.

there is a prophecy of Balaam recorded by Moses to this effect, 'there shall arise a star out of Jacob and a man shall rise up out of Israel' (Numbers 24:17)." (It is interesting to note that *The New English Bible* translates the passage: "A star shall come forth out of Jacob, a comet arise from Israel.")

Origen thereby contributed three important concepts to the legend of the star of Bethlehem. By repeating the Old Testament prophecy he implied a connection with the old tradition that the birth of prophets such as Abraham and Moses was heralded by a star, stressing the continuity between the two testaments. He clearly associated the star with a comet. And he stated unambiguously that comets could portend good.

John of Damascus, a Byzantine dogmatician whose writings were translated into Latin in the 12th century and who was one of the few Greek church fathers known to 12th- and 13th-century Europeans, related the star to a comet in his influential *Exact Exposition of the Orthodox Faith*, albeit less directly. "It often happens that comets arise. These...are not any of the stars that were made in the beginning, but are formed at the same time by divine command and again dissolved. And so not even the star which the Magi saw...is of the number of those that [were] made in the beginning. And this is evidently the case because sometimes its course was from east to west, and sometimes from north to south. At one moment it was hidden, and at the next it was revealed: which is quite out of harmony with the order and nature of the stars." That description surely connoted a comet to a reader.

The writings of Flavius Josephus, a Jewish historian of the first century A.D., support the association of the star with a comet. In *The Jewish War* Josephus mentioned in conjunction the portent of a broadsword-shaped star over Jerusalem and "a comet that remained a whole year." This was probably a confused reference to the apparition of Halley's Comet in A.D. 66, just before the outbreak of the Jewish revolt against Rome. Josephus listed other portents, among them a cow that gave birth to a lamb. Writers on Christian doctrine later read these details as an allegory of the star of Bethlehem announcing the birth of Christ (the lamb), whereas Jewish writers considered the passage to be prescient of the burning of the Temple in A.D. 70. (The apparition of A.D. 66 might well have influenced Matthew, whose Gospel was set down after the fall of Jerusalem.)

There was a very popular tradition in Giotto's own time linking the biblical star with a comet. In the widely read *Golden Legend* of 1275 (which has been suggested as a source for other details of

SCENES FROM INFANCY OF CHRIST are depicted on a page of an illuminated concordance of the four Evangelists executed in 1399 by the Paduan Jacopo Gradenigo. The painting contains two representations of the star of Bethlehem. Over the manger (*upper left*) it is a stationary star; above the Magi (*top center*) it is shown as a schematic comet, part angel and part star, with swooping rays that project forward as well as backward to form the tail.

ALBRECHT DÜRER included a luminous and dynamic comet, much less naturalistic than Giotto's, in his famous engraving *Melencolia I* of 1514. This is a detail from the upper left-hand corner of the complex print, which is interpreted as a "spiritual self-portrait" depicting the artist's melancholy at his inability to unite theory and practice. The comet and rainbow may symbolize the disease, insanity and miraculous events once associated with melancholics.

the Arena Chapel cycle but not up to now for this one) the Genoese theologian and chronicler Jacobus de Veragine described the star in a way that would have suggested a comet to his readers: "It was a star new created and made of God.... Fulgentius [a sixth-century defender of orthodoxy] saith: It differed from the other stars in three things. First, in situation, for it was not fixed in the firmament but it hung in the air nigh to the earth. Secondly, in clearness.... Thirdly, in moving, for it went alway before the kings." Jacobus also commented on the etymology of "epiphany," stating that it was derived from *epi* ("above") and *phanes* ("apparition") and thereby identifying the Epiphany (the adoration of Christ by the Magi) with the apparition of a divine star. "Magi" itself is derived from the Greek *magoi,* which in turn comes from a Persian word for sorcerers and astrologers, the wise men who studied the stars.

Language as well as literature, then, supported the close linking of the biblical star and the Adoration of the Magi with an irregular and spectacular apparition of a "new" and miraculous star, which was most likely to have been a comet. Giotto or his patron could have cited doctrinal authority for the representation of the star as a comet. It is also noteworthy that in Giotto's time there arose a more dispassionate—one might almost say scientific—attitude toward comets. One contemporary chronicler of the 1301 apparition concluded his description with an affirmation of the freedom of man's will as opposed to the power of the stars. Padua was a center of mathematics, the discipline that would eventually elevate astronomy from the superstitions of astrology; already in the late 13th century scholars had begun to observe the heavens assiduously. (Galileo was to hold a chair of mathematics at the University of Pad-

ua in the 16th century.) Giotto, then, worked in a philosophically progressive community that might well have encouraged the spirit of individualism and objectivity with which he regarded the comet and represented it in his fresco.

The revolutionary nature of Giotto's rendering is apparent when his comet is compared with representations of the star of Bethlehem by his contemporaries and successors. Giotto's chief rival of the period, the innovative but more Byzantine painter Duccio di Buoninsegna of Siena, executed an enormous altarpiece, the *Maestà,* for the Siena Cathedral between 1308 and 1311. In Duccio's *Nativity,* a panel from the predella, or base, of the altarpiece, he painted a static, symbolic eight-pointed star in the conventional iconographic tradition, with thin gilt rays pointing down to the Christ Child.

In about 1316, a decade after the com-

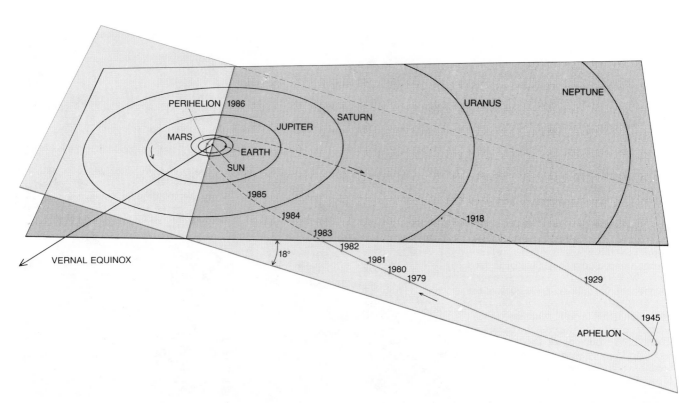

ORBIT OF HALLEY'S COMET (*color*) is an elongated ellipse with a major axis of about 35.6 astronomical units (1 A.U. is the mean distance from the earth to the sun) and a minor axis of about 9 A.U. The comet's motion is retrograde, that is, in a direction opposite to the motion of the planets around the sun; the plane of the orbit (*light color*) is inclined 18 degrees from the plane of the earth's orbit (*gray*). The comet will reach its perihelion, about .6 A.U. from the sun, on February 9, 1986. The colored dots show the position of the comet at perihelion and on the same date in some other years; the black dot shows where the earth will be when the comet reaches its perihelion.

pletion of the Arena Chapel cycle, a follower of Giotto painted a Nativity fresco in the Lower Church of San Francesco at Assisi, combining in the one scene an assortment of motifs from Giotto's Arena cycle. The anonymous artist confused his stellar images, painting a small conventionalized comet and a separate and more dominant stream of gold rays descending from heaven. (He painted an identical star-comet in the Adoration scene of the cycle, but it is damaged and barely visible.)

In contrast to Giotto's analytical depiction of Halley's Comet, the tiny Assisi comet is a stylized eight-pointed star like Duccio's, but with a wispy, linear tail; it seems to have been included as a timid afterthought mimicking the Arena comet. The artist either had never studied a comet apparition or had no desire to represent it convincingly; he wanted only to show the star as a symbol, relying on his knowledge that Giotto had painted it as a comet. Perhaps the separate rays from heaven implied theological criticism of Giotto's unorthodox depiction; by stressing the glorifying rays the Assisi artist may have meant to reinforce in a dogmatic way the divine nature of the event and the blessing conferred on it. Perhaps the less daring Assisi interpretation was required by the very public nature of this major pilgrim-

age church. The Arena Chapel, on the other hand, was private, providing Giotto with a more relaxed environment for both theological and artistic experimentation.

The advanced status of Giotto's portrait of the comet is accentuated by comparison with a much later work by another Paduan artist. In 1399 Jacopo Gradenigo executed an illuminated manuscript, a concordance of the Gospels. The star of Bethlehem over the Magi in his Infancy cycle is not a real comet but a symbolic representation, part angel and part star, culminating in a long, sweeping tail. It is highly schematic and has little to do with Giotto's naturalistic conception, even though it postdates the Arena Chapel fresco by almost a century. Illuminated manuscripts tend to reflect earlier innovations in more monumental painting. Gradenigo was not an innovator: the conservative nature of his art is evident in his adherence to the old device of "continuous narrative," with six temporally different scenes shown in the same illustration. Gradenigo's representation of the star as a comet, stylized though it is, demonstrates that by the end of the 14th century the comet interpretation of the star of Bethlehem had gained acceptance in Padua.

The comet did not supplant the more

usual interpretation of the star in European art in general, however. We do not even know if Giotto repeated it, because the Arena Chapel painting is the only extant depiction of the star of Bethlehem securely attributed to him. It seems futile to speculate about whether the idea of representing the star of Bethlehem as a comet was conceived by Giotto, by his patron Scrovegni or by a theological adviser. Apart from the question of iconographical innovation the significance of the *Adoration* comet lies in its striking naturalism, which surpasses even the comet depicted by Dürer in 1514 in his *Melencolia I*.

It is clear that by Giotto's time a substantial body of tradition associated the star of Bethlehem with a comet. Giotto drew on that tradition and on his own observation of Halley's Comet in 1301 when he painted his *Adoration,* and in doing so he depicted a celestial body that reflected naturalistic truth even as it heightened emotional intensity. Moreover, by painting a historical comet Giotto enhanced the contemporary impact of his depiction of the Adoration. He encouraged his viewers to identify with the biblical witnesses of the miraculous event of Christ's birth: they too had experienced a comet's spectacular apparition, in 1301.

SCIENTIFIC AMERICAN
Reports on the 1910
Apparition of Halley's Comet

2

HALLEY'S COMET.

BY DR. ALEXANDER W. ROBERTS.

Already in the pages of this magazine reference has been made to the approach of Halley's comet, the most important astronomical event of the years 1909 and 1910. Every seventy-five or seventy-six years this remarkable body completes its far-stretching and extremely elliptical orbit round the sun.

When the comet sweeps round the sun at its nearest approach, or perihelion, it passes within the earth's orbit; while at its farthest reach, or aphelion, it lies outside the confines of the solar system. Thus during the greater part of its long journey, for at least seventy-three years out of the seventy-five, or seventy-six, it is invisible in even the most powerful telescopes. It is then describing that portion of its path which lies outside the orbit of Jupiter. When, however, it is within the orbit of this planet, it is near enough to our earth to be visible in our evening or morning skies.

At first, on its sunward flight, it is discernible only as a faint telescopic object, but each day witnesses its increase in brightness, till at length it is visible to the naked eye, is, indeed, conspicuous enough to compel the gaze of even the indifferent beholder. As a rule it is easily seen by the unaided eye for some months.

After passing its brightest phase—which it will do this cycle in the first week in June, 1910—it rapidly decreases in brightness and is soon lost to view in even the largest telescopes.

It was last seen at the Cape Observatory in May, 1836, passing after that date into the far distances from which it came. But although it vanished from the sight of men, its onward track through space was known with as great accuracy, relatively, as sailors know the way of a ship over the trackless deep. And thus every lap of its vast orbit, over three thousand million miles distant, at its widest reach, from our earth, has been mapped out with the utmost care, and with assurance. Unseen for seventy-three years, it is yet as surely seen by those who make this branch of astronomy their care as if it shone brightly and continuously in our midnight sky. The invisible bonds of law have it in their inexorable hold, and from out the confines of that unbreakable leash it can never, never pass.

In the accompanying figure are given positions of the comet at various dates along the 1835-1910 cycle.

We have already said that the comet was last seen in May, 1836. It was then moving swiftly away from the sun, midway between the orbits of Mars and Jupiter. In the early days of 1837 it crossed the orbit of Jupiter. Jupiter himself was not very far away when the comet passed under his line of march. Slowing now down considerably, the year 1838 is well advanced before the region of Saturn's sway is reached. In six more years Halley's comet is as far distant as Uranus, and in twenty more years it is out beyond the farthest planet. And now, like a great, stately ship wearing in midocean, the comet slowly sweeps round in its orbit. Its long outward flight is spent, and the conquering homeward pull draws it sunward again. The year 1872 marked the comet's farthest distance, its aphelion; after this date its return journey begins.

At the opening of this century it was again within the orbit of Uranus. By the end of 1907 it had reached Saturn's orbit; and early this year it swept within the orbit of Jupiter.

On the first of June this year it was five hundred million miles distant from us, but rushing in at an ever-increasing speed. In June its velocity of approach was a million miles a day.

It will come nearest to our earth the first week of June, 1910, being then only twenty million miles distant from us—a hand-breadth in astronomical reckoning.

After this date it will move swiftly away from the earth, becoming daily more faint, till in the early days of 1911 it will disappear into the night, not to emerge again till the year 1985, when the most of those who read this article will have ceased to care about comets.

No small emulation is being witnessed between those observatories endowed with large telescopes, as to which one will be the first to pick up the returning voyager from far-distant shores. It is expected that this will be done in August or September of this year. The comet will then be a faint, nebulous star not far from Orion. But with regard to this matter of search, it may be said that it has already begun, chiefly by means of photography; it being thought that this auxiliary to science might make visible fainter objects than the eye can see.

In July of this year the comet held its course in the morning constellations, and was then badly placed for northern observers. It is so well placed, however, for southern observers that there is a hope that some observatory south of the line may have the

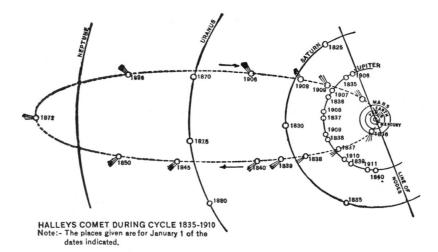

HALLEYS COMET DURING CYCLE 1835-1910
Note:- The places given are for January 1 of the
dates indicated.

good fortune to pick up the comet before the lengthening nights will enable northern observatories, armed with huge telescopes, readily to pick it up in the northern autumn. From September, 1909, to March, 1910, the comet will be well situated for observers all over the world, being then high up in the midnight sky.

By the month of April, 1910, it will have passed right round to the evening sky, and toward the end of the month it will be lost in the rays of the sun. During May it is, for a brief time, a morning star; but in June it has again stolen round to the west, and should, in that month, be an object of conspicuous brightness in the evening sky of southern latitudes. It will remain an evening star till October of that year, when once more it will pass behind the sun, and then will appear as a morning star. By this time, however, it will be rapidly decreasing in brightness and will soon be lost to view.

There are few more interesting chapters in the history of astronomical research than that which tells the tale of the tracing back through the long centuries of this remarkable comet. Chiefly through the devoted labors of a group of classical and Oriental scholars, we can trace back the appearances of Halley's comet through twenty centuries, surely a long enough period to give it a distinction among comets.

Some of these appearances are of unique interest. On the Bayeux tapestry there is a famous picture of the comet which appeared in 1066 and which William the Conqueror regarded as a herald of victory for his arms. Hind proved that this fateful star was none other than Halley's comet.

Then, again, there are many references in classical literature to the appearance of various "fearful stars," some of which can be claimed as returns of Halley's comet.

It is a marvel that the deductive minds of many of the ancient philosophers did not see a connection —as they did in eclipses—between regularly recurring phenomena. Possibly the fear which these supposed messengers of doom raised in the hearts of all, learned or unlearned, may have led the ancients to leave comets alone.

It is to Edmund Halley, the contemporary and friend of Newton, that we are indebted for lifting comets from the region of superstition to the calmer sphere of pure geometry. At Newton's request he undertook a thorough investigation of their movements, and into the laws which controlled these movements. The 1682 comet especially held his attention. His keen mind soon traced a connection between similar appearances separated by seventy-six years, and on working out the orbit of the comet, which will now forever be inseparably connected with his name, he boldly declared that it would again appear in 1758. It was seen for the first time on December 25th of that year by an amateur astronomer in Saxony.

Thus for ever the mystery which had gathered round comets was dispelled.

There are one or two very interesting questions connected with the return of this comet. Of these one is the constitution of these bodies. The usually accepted view is that they are composed of myriads of meteors at a very high temperature. But this does not explain their appearance thoroughly. It is indeed rather an imperfect explanation of the tail.

Then, do comets grow fainter each return? The belief prevails that they do. A comparison of the forthcoming appearance of Halley's comet in 1910 with the magnificent drawings made by Sir John Herschel of the 1835 appearance should do much to settle this question.

It is said that there are planets exterior to Neptune. Dr. See, of Mare Island Observatory, has even given them a local habitation and a name. If there are extra-Neptunian planets, then they should make their presence felt by perturbations of such comets as pass out beyond the orbit of Neptune. We have already referred to the near approach of Jupiter to Halley's comet in 1838. This approach would have the effect of pulling the comet back in its orbit, and thus by decreasing its centrifugal force bring it back more quickly again to the sun. Every planet circling round the sun tugs at the comet more or less, now hastening, now retarding its journey. Because of such "interferences" its path round the sun is a sinuous curve; and sometimes because of hindrances by the way, it takes seventy-seven years to return, while at other times it is incontinently pushed onward and its round is seventy-five years.

Thus if there are planets beyond Neptune they will make their presence felt in disturbing the comet as it passes its aphelion goal.

Enough has been said to indicate how much interest attaches to the appearance of this comet in 1910, and how eagerly its coming is being watched and waited for.

RETURN OF HALLEY'S COMET.

AN EPHEMERIS FOR THIS YEAR.

On September 11th, 1909, Prof. Max Wolf, at Heidelberg, detected the image of Halley's comet on a photograph taken (presumably) with the Bruce telescope of the Königstuhl Observatory. It is extremely satisfactory to learn that the comet was found in very close proximity to the place predicted in the ephemeris computed by Messrs. Cowell and Crommelin. Photographs of the region obtained at Greenwich with the 30-inch reflector on September 9th also show the comet, but this was not recognized until the definite news came from Prof. Max Wolf. The closeness of the predicted positions will be seen from the following comparison:

	R. A.	Decl.
Position at 14h. 7.3m. Königstuhl mean time	6h. 18m. 12s.	+ 17° 11'
Computed from Cowell and Crommelin's Ephemeris........... ....:	6h. 18m. 4s.	+ 17° 16'

Subsequently photographs were obtained at the Lick and Yerkes observatories, and from the positions thus available a corrected ephemeris has been prepared.— (A. C. D. Crommelin, Observatory, October, 1909, p. 400.)

EPHEMERIS FOR OBSERVATIONS OF HALLEY'S COMET AT 8:40 P. M. GREENWICH MEAN TIME.

	R A.	Decl.	Mag.
1909, November 1,.... ...	5h. 51m. 40s.	+ 16° 52' N	14.0
November 6.	5 42 33	16 49	13.7
November 11.	5 31 32	16 44	....
November 16.	5 18 33	16 38	13.2
November 21.	5 3 23	16 26	....
November 26.	4 46 13	16 13	12.7
December 1. ...	4 26 56	15 52	....
December 6.....	4 6 13	15 23	12.3
December 11. ...	3 44 24	14 45	....
December 16..... ..	3 22 19	14 4	11.9
December 21....	3 0 34	13 18	...
December 26.	2 40 11	12 28	11.7

REVISED ELEMENTS.

Perihelion passage .... 1910, April 20th
Longitude of ascending node. 57° 16' 12"
Node to perihelion......... 111 42 16
Inclination of orbit........ 162 12 42
Semi-major axis of ellipse.. 17.94527
Eccentricity 0.967281

Dr. Downing thinks that from the observed photographic magnitude of the comet it should be visible to the eye with a telescope of about 12 inches aperture. On October 15th its magnitude was about 14.5, its distance being about 230 million miles. At the beginning of November it will be almost in the continuation of the line through ξ and ν Orionis, and nearly north of the red star Betelgeuse (α Orionis). Of course, it is possible that the comet may be specially rich in blue and violet radiation, as was the case with Comet Morehouse, and not be easily visible to the eye at present even with considerable optical aid. Situated near the northern border of the constellation of Orion it will be excellently adapted for observation after midnight throughout the next two months. It is calculated to be approaching the earth at the rate of about 1,500,000 miles per day.

Mr. Crommelin finds that there is great probability that the comet will transit the sun's disk some time next spring (about May 18th), and this suggests the interesting problem of attempting to photograph it by means of the spectroheliograph. If we are fortunate enough to become familiar with its spectrum in the near future, it will be possible to suggest what wavelength should be transmitted by the secondary slit of the instrument, and it is hoped that by the relative intensification of the special cometary radiation some indication of the nucleus and coma may be photographed. A detailed ephemeris for this will be furnished later.—Knowledge and Scientific News.

COULD THE EARTH COLLIDE WITH A COMET?

ON May 18th next the earth will be plunged into the tail of Halley's comet, and the head of that body will be but 15,000,000 miles away. It is but natural that a thinking man should ask: Is there a possibility that the earth may encounter a comet and thus come to a frightful end?

Curiously enough, it was Halley himself who first pointed out the possibility. Whiston, Newton's successor in the Lucasian chair of mathematics at Cambridge, was so alarmed at "a chariot of fire" which flared up in his day, that Halley was prompted to look closely into its movements. His work led to the startling result that the comet, when passing through the descending node, had approached the earth's path within a semi-diameter of the earth. Naturally, Halley wondered what would have happened had the earth and the comet been actually so close together in their respective orbits. Assuming the comet's mass to have been comparable with that of the earth (an assumption which we now know to have been utterly beyond reason) he concluded that their mutual gravitation would have

caused a change in the position of the earth in its orbit, and consequently in the length of a year. This train of thought led him to consider what the result of an actual collision would have been, and he concludes that "if so large a body with so rapid a motion were to strike the earth—a thing by no means impossible—the shock might reduce this beautiful world to its original chaos."

Hence Halley not only dispelled the superstition and the terror which once followed in a comet's wake, but also pointed out a possibility which the superstitious Dark Ages had ever dreamed of. It seemed to Halley not improbable that the earth had at some remote period been struck by a comet which, coming upon it obliquely, had changed the position of the axis of rotation, the north pole having originally, he thought, been at a point not far from Hudson's Bay. The more recent investigations of Kelvin and Sir George Darwin completely upset any such theory.

Since Halley's time the chance of a collision between the earth and a comet has engaged the attention of many astronomical mathematicians. Laplace, for example, painted the possibility of a collision with the earth so vividly that he startled his day and generation.

He drew a picture of a comet whose mass was such that a tidal wave some 13,000 or 14,000 feet high inundated the world, with the result that only the higher peaks of the Himalayas and the Alps protruded. Lalande created a panic by a similar consideration of the subject in a paper which was intended for presentation before the Academy of Sciences, but which was not read. Such was the popular excitement, that he felt himself constrained to allay the public fears as well as he could in a soothing article published in the Gazette de France. The masses assumed by both Laplace and Lalande are so preposterous that their theories are no longer seriously considered by any sane astronomer.

Since the day of Laplace and Lalande there have been several comet "scares." Biela's comet crossed the earth's orbit on October 29th, 1832. When that fact was announced, Europe was in a ferment. The orbit of the earth was confused with the earth itself. Such was the popular excitement, that Arago took it upon himself to compute the possibilities of a collision. He pointed out that the earth did not reach the exact spot where the comet had intersected the earth's orbit until a month later, on November 30th, on which date the comet was 60,000,000 miles away. Incidentally he pointed out that a collision was always happily remote. He thought that the chances of a meeting were about one in 281,000,000. Babinet, on the other hand, thought that a collision was likely to take place once in about 15,000,000 years. More recently the entire problem has been considered by Prof. W. H. Pickering of Harvard. By a collision he understands, first, that any part of the earth strikes any part of the comet's head; second, that any part of the earth strikes the most condensed point in the head (the core) as distinguished from the larger nucleus. What the average size of a visible comet's head may be, we have no means of knowing. Young estimates that for a telescopic comet it averages from 40,000 to 100,000 miles in diameter. The head of the great comet of 1811 was 1,200,000 miles; that of Holme's comet in 1892, 700,000 miles; and that of naked-eye comets generally over 100,000 miles.

In the last half of the last century 121 comets, including returns, penetrated the sphere of the earth's orbit. From this Prof. Pickering infers that we should expect to be struck by the core of a visible comet once in about 40,000,000 years, and by some portion of the head once in 4,000,000 years. Since comets' orbits are more thickly distributed near the ecliptic than in other regions of the sphere, the collisions would occur rather more frequently than this, but hardly as often as once in 2,000,000 years; and since it has been estimated that animal life has existed upon the earth for about 100,-000,000 years, a considerable number of collisions, perhaps as many as fifty, must have taken place during that interval, in Prof. Pickering's opinion, evidently without producing any very serious results.

The old notions of the tidal effects of comets were based upon an erroneous conception of cometary masses. It seems astonishing that a man of Laplace's wonderful mathematical powers should not have concluded that a body like a comet, which can sweep through the entire solar system without deranging a single one of its members, must have a mass so small that it cannot appreciably affect the waters of the earth. As it is, comets are more likely to be captured by planets (witness the comet families of Jupiter and Saturn) than to derange a member of the solar system or to produce tidal effects.

The plunging of the earth in the tail of Halley's comet naturally causes many to wonder what will be the effect upon the inhabitants of the earth. Similar passages occurred in 1819 and 1861, but no one was the wiser until long after. Some astronomers claimed to have noticed auroral glares and meteoric displays at the time, but whether these were really associated with the comet or not cannot definitely be stated. At all events, it may be safely held that on May 18th next none of us will be aware of the fact that we are literally breathing the tail of Halley's comet. From this it may well be inferred that the wild tales of the possible effects of poisonous gases, tales for which the newspapers are very largely responsible, are utterly without foundation. It is true that a comet's tail is composed of poisonous and asphyxiating hydrocarbon vapors and of cyanogen; but it is also true that the actual amount of toxic vapor is so small that when the earth is brushed by the tail of Halley's comet, the composition of the atmosphere will not be so affected that a chemist could detect it. Flammarion has drawn a vivid picture in his "La Fin du Monde" of the possible effect of passing through a tail highly charged with vapors. He has shown us terrified humanity gasping for breath in its death struggle with carbon monoxide gas, killed off with merciful swiftness by cyanogen, and dancing joyously to an anæsthetic death, produced by the conversion of the atmosphere into nitrous oxide or dentist's "laughing gas." No one of any common sense should be alarmed by these nightmares, particularly when it is considered that so diaphanously thin is a comet's tail, that stars can be seen through it without diminution in brightness.

CONDENSED FACTS ABOUT HALLEY'S COMET.

BY H. W. GRIGGS.

A few facts presented in a condensed form may possibly interest the readers of the SCIENTIFIC AMERICAN who wish to follow the course of Halley's comet in the heavens during its present appearance.

The last perihelion passage occurred on November 15th, 1835. The present perihelion passage will occur on April 20th, 1910. The perihelion distance will be 0.587, and the aphelion distance will be 35.30. The eccentricity is 0.967, the longitude of ascending node is 57 deg. 16 min., the node and apsis angle is 111 deg. 47 min., the inclination of the orbit is (162 deg. 12 min. +) 17 deg. 48 min. —, the longitude of perihelion is 305 deg. + and the motion is retrograde, in other words opposite to that of the planet. The diameter of the nucleus cannot of course be stated with anything like accuracy at the present time, but it is not likely to exceed 120,000 miles.

At the end of February Prof. Barnard of Yerkes Observatory estimated the tail to be 14,000,000 miles long. Just before and after perihelion passage the tail will be at least that long, and probably longer. The comet is fast approaching its perihelion point, or point nearest the sun, where, as we have said, it is due to arrive on April 20th. During the months of February and March, the earth and the comet are racing on practically parallel orbits, 170,000,000 miles apart on opposite sides of the sun.

The comet first crossed the earth's orbit about March 10th at a point where the earth will arrive at the middle of next October, but far above where the earth will be, so to speak, for it will be some 10,000,000 miles above the plane of the ecliptic. In April the comet will emerge from behind the sun, and will become visible to the naked eye in the eastern sky before sunrise.

On April 20th, when the comet will swing around the sun, it will be 57,000,000 miles away from the sun. Its velocity will be 26 miles a second. The earth travels at about 19 miles a second. On May 2nd the comet will traverse the orbit of Venus, some 6,000,000 miles above the planet. In other words, an astronomer on Venus would find the comet a far more impressive spectacle than a terrestrial astronomer. As it rushes on, Halley's comet will pass between the earth and sun close to its ascending node. On May 18th the earth will be about 13,000,000 miles away from the nucleus or head, as against 5,000,000 miles in 1835. Moreover, on May 18th the earth will be enveloped in the comet's tail for a few hours. A few days later the comet will be visible in the western sky after sunset with a 15 deg. or 20 deg. splendor. After that it will speed away from the solar system. The last glimpse of it with the naked eye will be obtained probably at the end of June. It will not reappear for seventy-five years.

Halley's comet is noteworthy because it was the first comet for which an orbit was plotted and a time table calculated. It has a history more or less identified with the history of human thought and civilization. The superstitious dread with which it was regarded in medieval and ancient times swayed many a monarch. It was instrumental in forming the policies of Louis le Debonnaire in 837. It blazed in the sky when the Turks threatened to overrun Europe in 1456, and when the Reformation was at its height in 1531. It struck terror to the Saxons under Harold in 1066, when they were conquered by William of Normandy. This fear of the middle ages was dispelled only when Halley made his great prediction in 1682 that the comet would return in 1758, a prediction which was verified after the great astronomer was in his grave.

A comet which has reappeared regularly for over two thousand years must be composed of fairly enduring stuff. Just what its composition may be, the present reappearance will for the first time enable us to tell, for in 1835 the spectroscope was not invented, nor astronomical photography perfected.

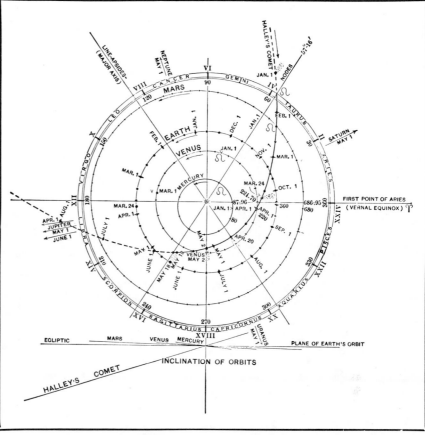

HALLEY'S COMET AND THE EARTH.

Dots on the orbits show positions of planets and comet every 10 days. Positions for January 1st, 1910, are shown thus : "Jan, 1." The ascending nodes, or points where the orbits first cross the ecliptic, are shown thus : ☊. Dotted portions of the orbits indicate the part below the ecliptic. The outer circle shows the signs of the zodiac. The celestial (heliocentric) longitude and the right ascension are indicated in hours. The Inclination Diagram shows the great angle of the comet's orbit.

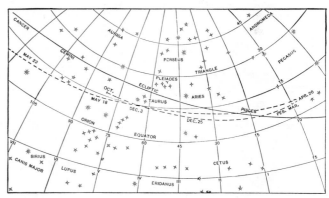

THE APPARENT PATH OF HALLEY'S COMET THROUGH THE HEAVENS.

SCIENTIFIC AMERICAN

[Entered at the Post Office of New York, N. Y., as Second Class Matter. Copyright, 1910, by Munn & Co., Inc.]

A POPULAR ILLUSTRATED WEEKLY OF THE WORLD'S PROGRESS

Vol. CII.—No. 16.]
Established 1845.]

NEW YORK, APRIL 16, 1910.

[10 CENTS A COPY.
[$3.00 A YEAR.

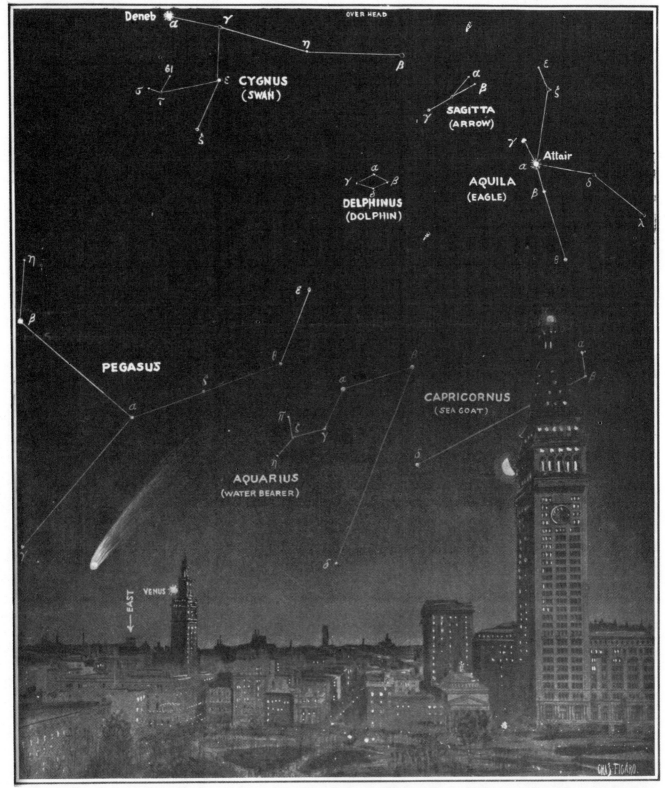

Halley's comet is now visible to the naked eye in the eastern skies, just before dawn. Its appearance on May 2nd, one hour before sunrise, is here depicted.

HALLEY'S COMET AT ITS BRIGHTEST.—[See page 317.]

HALLEY'S COMET AT ITS BRIGHTEST.

BY HENRY NORRIS RUSSELL, PH.D., PROFESSOR OF ASTRONOMY AT PRINCETON UNIVERSITY.

It may have seemed remarkable to many people that so long a time has elapsed since the first observation of Halley's comet at its present return, and yet it has not shown itself at all to ordinary eyes. The accompanying illustration (Fig. 1) will help to explain this. When first detected last September, with very powerful telescopic aid, it was far beyond the limits of our diagram, at twice the distance of Mars from the sun, and nearly as remote from the earth. At first the two bodies approached each other rapidly; but before the end of the year our planet crossed the line joining the comet with the sun, and by January 1st, as the figure shows, we were moving almost straight away from it. During the early part of the year the earth and comet passed on opposite sides of the sun, so that it was lost to our view early in March.

About the time that this is printed it will come into sight again, on the other side of the sun, rising before daybreak. But now its path has curved so that it is coming toward us—almost directly, if we take our motion into account as well as its own. It therefore seems to stand almost still among the stars, while growing steadily larger and brighter, so that any one might tell by its mere changes in appearance that it was approaching us rapidly.

Finally, about the middle of May the comet will apparently approach the sun again, and on the 18th it will pass in front of him, literally between us and the sun, transiting the latter's disk. If at this time its tail is more than fifteen million miles in length we will pass through it, as the figure shows.*

The comet's closest approach to us comes two days later, on May 20th, when it is but fourteen million miles away. For a few days following this it will be splendidly visible in the evening sky, and then it will fade gradually as it recedes from us.

It is clear from the diagram that this apparition of the comet is an exceptionally favorable one, for it passes the earth almost at the point where their orbits come nearest to one another. If it had returned only three weeks earlier, it would have come as near as possible—only seven millions of miles—but at this time it would have been directly south of the earth, astronomically speaking, almost over our south pole, and quite invisible from northern latitudes. It therefore appears that the present conditions are almost ideally favorable for observers placed, as we are, north of the equator.

The illustration on the first page shows better than any verbal description where to look for the comet in the morning sky in New York. The moon and Venus are shown in the positions which they will occupy about May 1st, when, on the whole, the comet can be seen to the best advantage. At an earlier date, Venus was higher in the sky, compared with the comet. There was less trouble then from moonlight, but the comet did not rise so early—about 4 A. M. on April 15th as against 3 A. M., on the later date.

The comet's brightness when it appears in the evening sky about May 20th will be sufficient to render any finding diagram unnecessary. It will only be needful to look toward the west half an hour or more before the comet sets, which it does at 8:20 P. M. on the 20th, 9:15 on the 21st, and 9:55 on the 22nd, after which it will be clearly visible until after 10 P. M.

* Prof. Barnard has informed us that the tail was 14,000,000 miles long on February 27th, from which it may well be inferred that it is much longer than 15,000,000 miles now.—ED.

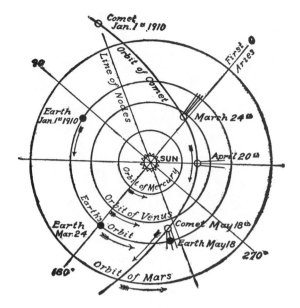

Fig. 1.—RELATIVE POSITIONS OF HALLEY'S COMET, THE EARTH, AND THE SUN.

Our other illustrations, which appear here through the courtesy of Profs. Frost and Barnard of the Yerkes Observatory, show the appearance and character of the comet earlier in its apparition. Fig. 2 illustrates its extreme faintness at the time of its rediscovery (which was announced by Prof. Wolf of Heidelberg less than a week before the earliest of the four photographs here shown was taken) while it was still 300 million miles distant, both from the earth and from the sun. On any one plate it is difficult, if not impossible, to distinguish the comet from the multitude of faint stars around it, but on comparing the four (which show exactly the same region of the sky) it is easy to see that the stars are the same in all, while the comet is "here to-day and gone to-morrow."

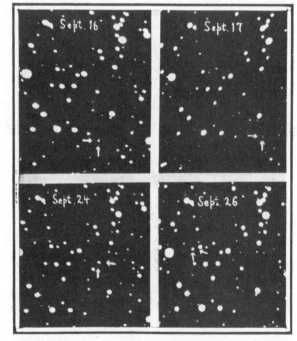

Fig. 2.—Halley's comet at its reappearance in 1909.

From photographs taken by Mr. Lee with the two-foot reflector of the Yerkes Observatory. These four photographs represent the same portion of the sky. The arrows point to the comet which appears like a faint star but moves from night to night.

Fig. 3.—Halley's comet on February 3rd, 1910.

From a photograph taken at the Yerkes Observatory by Prof. Barnard. As the instrument was kept pointed at the comet during the exposure the stars appear as short streaks. The actual length of the comet's tail is about five million miles.

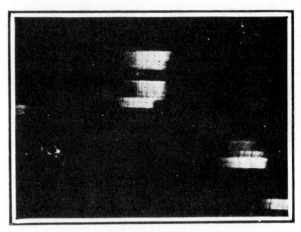

Fig. 4. Spectrum of Halley's comet.

Photographed at the Yerkes Observatory by Prof. Frost, January 14th, 1910. The spectrum of the comet is in the middle between the two brightest star-spectra. See description in text. The blue end of the spectrum is on the right; the ultra-violet on the left.

With the great Yerkes telescope (which gives far smaller and sharper images of the stars than can be reproduced on any known photographic plate) the comet was even at this time quite different from the stars in appearance; in Prof. Barnard's words, "a fleck of light surrounded by a faint nebulosity" with no definite boundary. His measures, made on several nights, show that its actual diameter was about 12,500 miles.

Our second illustration, from a photograph taken when the comet was 143 million miles from the sun, and 162 million from us, shows it already well advanced in the changes which invariably accompany the approach of any considerable comet to its perihelion. The head of the comet has become larger—not merely in apparent size, owing to its approach to us, but actually in miles, while a faint slender tail, pointed away from the sun, makes its appearance.

As Fig. 1 shows, the tail, which extends directly away from the sun, was at this time also nearly in line behind the head, as seen from the earth, so that its actual length must have been much greater than it appears to be—about five million miles, according to Prof. Barnard.

This considerable development of the tail, while the comet was still at two and one-half times its least distance from the sun, makes it probable that at and after the perihelion passage, on April 20th, it will be much longer, probably long enough to envelop the earth as it sweeps past.

Our third illustration shows the spectrum of the comet photographed on January 14th, when it was about 170 million miles from the sun.

In taking such a photograph, a prism is placed in front of the camera. The light of a star is thus drawn out into a line which, by letting it trail on the plate, is broadened into a band, crossed by the dark lines which tell us what absorbing gases exist in the star's atmosphere. Most of the objects on the plate are the spectra of stars near the comet obtained in this way. The comet's spectrum is near the middle, between the two very broad and conspicuous "comparison spectra," which were produced by supplementary exposures on some bright star, and serve as reference marks to find the position of the lines in the spectrum of the comet itself. The latter, unlike that of the stars. consists mainly of bright bands or lines, three of which are

conspicuous. The brightest of these, as is shown by comparison with the hydrogen lines of the comparison spectrum, is the so-called "cyanogen band" at the extreme violet end of the visual spectrum. The others are probably, as in the case of other comets, also due to gaseous compounds of carbon.

Between these bright bands can be seen a faint continuous spectrum, due to reflected sunlight.

When the comet first appeared, the photographs made at the Lick Observatory showed this continuous spectrum alone. At that time it must have been shining entirely by reflected light; but by the date of our illustration it had already begun to be self-luminous. This is corroborated by the fact that its brightness increased much more rapidly than could be explained by the mere increase in the amount of reflected light, due to its approach to the sun and to us.

This intrinsic light of the comet, as its spectrum shows, is given off by luminous gas; but we do not yet know what makes this gas shine. It can hardly be high temperature, for the comet had just come from the depths of interplanetary space, and did not yet receive nearly as much heat from the sun as the earth does. It must, however, be due to some kind of solar action, for it increases very rapidly as a comet approaches the sun. We can reproduce the same spectrum in the laboratory by passing an electrical discharge through a vacuum tube containing compounds of carbon and nitrogen at very low pressure.

It is of special interest that, even if the carbon compounds form but a small percentage of the gas in the tube, their spectrum becomes relatively prominent when the pressure is made very small, say 1/100,000 of that of ordinary air. It may be, therefore, that at the lowest pressures, carbon compounds have an exceptional capacity for shining; and it would be unsafe to conclude that they are the principal gaseous constituents of the comet, because they give off almost all the light.

It may be added that the "cyanogen" bands in the spectrum are produced, not only by the poisonous gas of that name, but in all cases when carbon and nitrogen are together under electrical excitement. For example, they are very strong in the spectrum of an ordinary arc-light, where the nitrogen comes from the arc, and the carbon from the terminals. It would be about as reasonable to conclude that an arc-light was poisonous, after looking at it through a spectroscope from a distance, as to make the same deduction about a comet.

Whatever may be the origin of this intrinsic light of comets, it is responsible for most of the phenomena which make them of general interest, for almost all the light of the tail, as well as of the head of a bright comet is of this kind. If Halley's comet shone by reflected sunlight alone it would be barely visible to the naked eye, even under the most favorable circumstances.

Actually, owing to its intrinsic light, it has been a conspicuous object at every return for the last 2,000 years. The only gap in the record—in A. D. 912—has lately been filled by the discovery of unmistakable references in old Japanese chronicles.

The actual quantity of matter composing it must, however, be very small as compared with the more familiar heavenly bodies. It is possible to form a rough guess as to its amount by considering the amount of light which it reflects when it is not shining on its own account. From the estimates of magnitude made last September, it appears that a single body only a little over 30 miles in diameter at the distance of the comet would have sent us as much reflected light, provided that its reflecting power was equal to that of the moon, which is lower than that of most of the planets.

It is, therefore, clear that the comet must be composed of separate particles widely separated. The whole cross-section of the comet (12,500 miles in diameter) is about 120 million square miles; while the total area of all the reflecting particles, according to the above estimate, is about 1,000 square miles. A ray of sunlight falling on it has therefore less than one chance in 100,000 of being stopped, and all the rest of getting through some empty space. It is no wonder that comets are transparent, and that stars can be seen through them! If we only knew how big these particles were, we could now estimate their number and their total mass. But here we are quite in the dark. As the light of the comet seems uniformly diffused and it shows no signs of resolution into points of light, the number of particles composing it must at least be counted by thousands. Their average diameter must therefore be less than a mile, though they may vary enormously in size. If all gathered into one compact group, they could at most hardly exceed in bulk the satellites of Mars or the smallest of the asteroids.

But how much smaller than this limit their actual dimensions may be, we do not know. If, purely for illustration, we suppose that they average an inch across, there would be some five or six million millions of them. This sounds like an enormous number, but if we calculate the bulk of the comet, we find that there would be only five or six particles per cubic mile of space, on the average, inside it. Near the center they would doubtless be more closely packed, and more thinly toward the outer parts of the comet. The combined bulk of all these particles would be about 80 million cubic yards—a large amount from the engineering standpoint, but not equal to the quantity of water which falls within the limits of the smallest State in the Union during a heavy rainstorm.

This may serve to give us some idea of the extreme tenuity of the comet as a whole. If we took a space as big as the comet, that is, half as much again in diameter as the earth, and sowed ordinary golf balls through it at the rate of two or three per cubic mile, leaving the intervening space absolutely vacant, we would get something that would look quite as bright as Halley's comet—if put alongside it when it first appeared.

The gaseous matter, which gives most of the light at perihelion, probably oozes out of the solid particles as these grow warm under the sun's heat when they approach it. As the gas becomes luminous under solar action, the brightness of the comet increases, and its outer regions, originally invisible because the number of reflecting particles was too small to influence our eyes, gradually come into view.

Some of this is repelled from the head of the comet, by little-known forces, and driven away from the sun by the action of the sunlight, which, as is well known, exerts a force of repulsion which, if a particle is exceedingly small, as are the gaseous molecules, is stronger than the attraction of the sun.

Thus arises the long and magnificent tail, which, like the smoke-trail of a steamer at sea, is ever being renewed at one end and fading away at the other, even though it seems to accompany the comet in its journey.

As the comet recedes from the sun, much of this gaseous matter has thus been lost, never to be regained. Some of the remainder probably condenses round the solid particles when they become cold, and some escapes into space.

The comet is thus gradually losing its substance, and in the course of ages it may be deprived of all its tail-forming material, and lose its former glory. This seems to have actually happened to some of the short-period comets, one at least of which has disappeared altogether.

Halley's comet is perhaps preserved from such a fate by the longer interval between its returns to the region near the sun, where its activity takes place. It may be, too, that it has more of the right sort of material to spare for a tail. But the time may come when most of this is lost, and its successive appearances may gradually lose those impressive features which have so long inspired awe and wonder in the hearts of mankind, and dwindle at last into something which the professional astronomer alone will be interested in watching.

THE HEAVENS IN MAY

BY HENRY NORRIS RUSSELL

IT is seldom that so much of interest to the amateur astronomer happens in a single month as in the one which is just before us.

First and foremost, of course, is the return of Halley's comet to the position where it is seen to the best advantage. Early in the month it is favorably placed for observation before daybreak; on the 18th it passes directly between us and the sun; and later it appears to even greater advantage in the evening sky.

At the beginning of May the comet is about 74 million miles away, but it approaches us rapidly, its distance diminishing to 41 million miles on the 10th, and 27 million on the 14th. As it was about at the limit of visibility to the naked eye on April 12th, while still 135 million miles from us, it is now a fairly conspicuous object.

The planet Venus is fortunately near by and serves as an excellent "pointer" to the comet. Anyone, however little familiar with the heavens, can easily find the latter by observing the following directions:

Choose a window from which the eastern sky is visible clear down to the horizon. Rise about 3:15 A. M. and look due east. The very bright starlike object, low down in the sky, is Venus. The comet is to the left of this and a little higher up at a distance about as great as the length of the bowl of the Great Dipper. It will probably be rather fainter than the four stars, forming a great square, which lie above and to the left of Venus, about twice as far away as the comet.

These directions hold good from May 1st to May 12th. On the 14th the comet will be on a level with Venus, and a little farther to the left. On the 16th it will be much lower than the planet and about 20 deg. to the left. After this the comet, or at least its head, can hardly be seen clear of the morning twilight.

It will be very interesting to watch the comet grow larger and brighter night by night as it comes nearer to us. How long its tail will be it is impossible to predict. The best time to see this, however, will in any case be from the 7th onward, when the moon is out of the way and the sky dark. The comet will be larger and brighter, too, at this time than previously.

Even after the head gets too near the sun to be seen, the tail may be observable in the mornings of the 17th and 18th extending upward and to the right from the eastern horizon, perhaps broad and fan-shaped, from the effects of perspective, since the end of it will be much nearer us than the head.

On the evening of the 18th or morning of the 19th (according to the observer's longitude) the comet passes between us and the sun, and the earth will be enveloped in its tail if the latter is long enough (over 15 million miles).

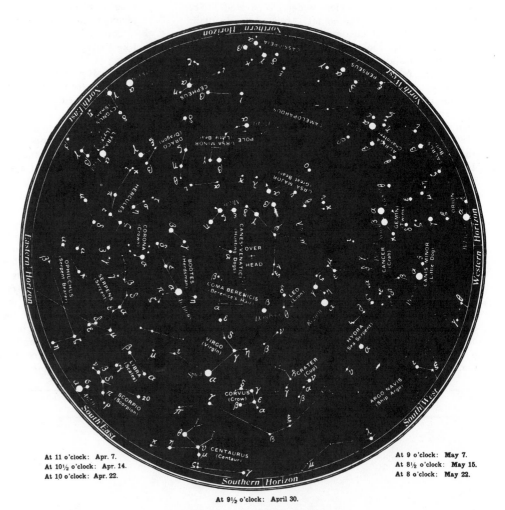

At 11 o'clock: Apr. 7.
At 10½ o'clock: Apr. 14.
At 10 o'clock: Apr. 22.

At 9 o'clock: May 7.
At 8½ o'clock: May 15.
At 8 o'clock: May 22.

At 9½ o'clock: April 30.

NIGHT SKY: APRIL AND MAY

If this evening is clear it will be of great interest and importance to look for illumination of the sky. In the early evening, just after sunset, the comet's tail will be in the east, but a few hours later it will have passed over toward the west. If, as is sometimes supposed, the tail is a hollow cone of light, there will be two times at which the sky in general is comparatively brightly illuminated, separated by an interval while we are in the darker center of the tail. Meanwhile, observers on the opposite side of our planet will have the rare privilege of seeing the sun through the comet's head. Only the extreme western portion of the United States is included in this favored region, but as the comet enters upon the sun's disk at 6:22 P. M. by Pacific standard time and remains on it till 7:22, the transit will be visible all along the coast. The comet passes almost squarely across the center of the sun from west to east.

Paradoxical as it may seem, it is probable that the ordinary observer, even with a small telescope and dark glasses, will not be able to detect even the slightest trace of the comet's passage. With powerful instruments the nucleus, if solid, might be seen as a dark speck against the sun, if it is over 50 miles in diameter; but it is improbable that it is anything like so large, for, as has already been stated in these columns, the whole amount of light reflected from the comet, when remote from the sun, is no more than a single mass 30 miles across would send us.

It is possible, too, that the absorption of the gases composing the envelopes of the head and the tail may be detected by means of the spectroscope, and as we will be looking through the tail lengthwise, nearly 15 million miles of it will be there to exert any possible effect on the sun's light. Even so, it will not be surprising to many astronomers if nothing unusual is detected.

Such negative results will however be scientifically valuable, since they will enable us to say that the materials composing the comet do not exceed certain limits of mass or density.

Transits of comets across the sun are very rare. The most remarkable previous instance is that of the great comet of 1882, which, though so bright that it could be seen close to the sun in broad daylight with the naked eye, vanished completely when in front of the sun's disk, showing that it was practically perfectly transparent.

On the evening of the 19th we may perhaps already see the comet's tail in the evening sky, though its head will set while the twilight is still very strong. On the 20th, however, it will be visible till about 9 P. M., on the 22nd till 10:20, and on the 25th and afterward until after 11 P. M.

On the 21st the comet's head will be close to the star γ Geminorum; on the 23rd about 10 deg. above Procyon; and on the 25th near ϵ and ζ Hydrae. Fuller details will be given later.

The Transit of Halley's Comet.

The transit of Halley's comet and the expected immersion of the earth in the tail of that historic body have proven once more what may happen to the best-laid plans of mathematicians. The transit undoubtedly occurred, but whether or not the earth really encountered the tail seems to be a matter of considerable doubt. When the night of May 18th came, and the scientific world was all agog, the tail was so curved that the passage of the earth through it seemed only remotely possible. On the morning of the 20th a broad band of light that stretched along the horizon for a distance varying from 120 to 160 degrees proclaimed indubitably that the earth was still untouched, and that contrary to expectations, the comet was still in the east. Prof. W. W. Campbell, of the Lick Observatory, saw the comet visibly in the eastern sky. According to him, the tail was at least 140 degrees long and lagged far behind the radius vector. Because of the angle of 18-odd degrees which separates the earth's orbit from that of the comet, the curvature of the tail, to which this extraordinary misadventure may probably be traced, probably prevented the earth from coming in contact with it.

All the scientific expeditions which have been sent out to various parts of the earth will probably come back with nothing to report. Some of these scientific parties must have proceeded to their destinations at considerable expense. Thus Prof. Birkeland, of the University of Christiania, went to Kaafjord, in the northern part of Norway, for the purpose of noting whatever electrical and magnetic effects might be attributed to the comet's tail, and particularly to observe the relation of the aurora borealis to the comet. He has his own theory that the particles of a comet's tail are so highly electrified that they may, in some way, affect the aurora borealis. The failure of his expedition, which it is almost safe to assert, still leaves that theory unproved.

Similarly unsuccessful must be all the elaborate preparations made by meteorologists, navigators and physicists. The more important meteorological stations of the world sent up sounding balloons at frequent intervals on the 18th and 19th of May, for the express purpose of bringing down from the upper strata of the atmosphere some record of unusual happenings which might safely be attributed to the influence of the comet. All this labor is now in vain. Similarly, it is very unlikely that the instructions sent forth by the United States Hydrographic Office to wireless operators, charging them to note any curious and unusual effects on their instruments, will prove barren.

The expedition which was sent to the Hawaiian Islands by the Astronomical and Astrophysical Society of America for the purpose of observing the transit cables a preliminary report of complete inability to note any transit whatever. This was more or less expected. In 1882 a transit occurred which was fortunately observed by Mr. Finlay at the Cape of Good Hope. The comet of 1882 was followed by him "continuously right into the boiling of the limb." No sooner had it touched it, than it vanished as if destroyed. So sudden was the disappearance, t h a t the comet was at first believed to have passed behind the sun. As a matter of fact, the observers at the Cape had witnessed a genuine transit. The experience of the observers at the Hawaiian Islands with Halley's comet seems to have been exactly similar. On the whole, this apparent failure to observe the creeping of a black speck across the face of the sun may be deemed a confirmation of our present theories that the bulk of a comet is much too flimsy to be detected in the blinding glare of our central luminary.

Although the passage of the earth through the tail of Halley's comet turned out to be an extraordinary disappointment, it is unfair t o charge o u r mathematical astronomers w i t h incompetence. A comet's tail is so capricious, so fluctuating a structure, it changes with s u c h startling rapidity, that the predictions of any astronomer with regard to its behavior must always be stated with some reserve.

T h e tail of Halley's comet has conducted itself in a most whimsical fashion. In the middle of February, it was some fifteen million miles long. In April, it seemed to have vanished entirely.

Then it grew again, until finally it attained a length that has been variously placed at twenty to forty million miles. It seems to have split longitudinally into three more or less well defined parts. When we consider that Morehouse's comet of 1908 (comet C, 1908) exhibited some extraordinary changes; that it repeatedly formed tails, which were discarded to drift out bodily into space, until they finally melted away; that in several cases tails were twisted or corkscrew shaped, as if they had gone out in a more or less spiral form; that areas of material connected with the tail would become visible at some distance from the head, where apparently no supply had reached it from the nucleus; that several times the matter of the tail was accelerated perpendicularly to its length; and that at one time the entire tail was thrown forward and violently curved perpendicularly to the radius vector in the general direction of the tail's sweep through space (a peculiarity opposed to the laws of gravitation), it is evident that a comet presents important problems for the future astronomer to solve. It is no wonder that Halley's comet should have disappointed us.

THE HEAVENS IN JULY

BY HENRY NORRIS RUSSELL, PH.D.

LEST we have access to powerful telescopes, the moonless evenings at the beginning of July afford the last opportunity to see Halley's comet that most of us will ever enjoy. Though 150 million miles from us, and 140 million from the sun, at the beginning of the month, it should still be visible with the aid of a field glass, and perhaps faintly to the naked eye. It is then in the southern extremity of Leo in R. A. 10 h. 48 m. and declination 2 deg. 50 min. south—three or four degrees above the intersection of lines drawn through the stars β and ε Virginis, and α and ε Leonis— all of which are shown on our map—and sets at about 10 p. m. With the telescope it will probably still show some traces of a tail.

After the 7th or 8th the light of the new moon will drown the comet out completely, and a fortnight later, when we can once more see our departing visitor on a dark sky, it will be in all probability too faint to be detected without telescopic aid. By the end of the month (when it sets at 8:30 p. m.) the twilight will interfere with further observations, and nothing more will be seen of it till the end of autumn, when it may be picked up in the morning sky with powerful telescopes, under conditions of brightness resembling those shortly after its rediscovery a year before.

The feature of its close approach to the earth which excited most discussion was undoubtedly the marked curvature of its tail, which caused it to lag so far behind the prolongation of the line joining the sun and comet that it did not graze the earth until more than thirty-six hours after the head had passed between us and the sun.

To be sure, the tail looked straight, but that was because we saw it from a point in or very near the plane of its curvature, like a piece of a hoop seen edgewise.

There was no reason, however, for anyone to be surprised at this behavior, for the tails of comets are usually curved; indeed, they must always be more or less so, as the following considerations will show. The tail consists of fine particles of gas and dust, detached from the head, and repelled by the sun. This repulsion steadily increases the speed with which they are moving away from the sun; but their rate of motion at right angles to the line joining them to the sun is unaltered by the repulsion and remains the same as before.

The line drawn from the sun through the comet, however, sweeps over wider and wider arcs at increasing distances from the sun. If, therefore, we consider a tail particle which has already receded

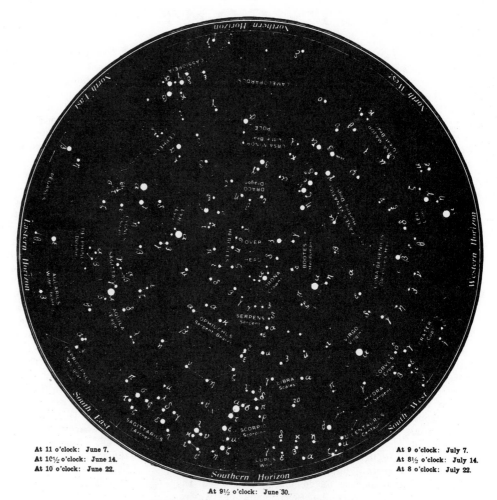

At 11 o'clock: June 7.
At 10½ o'clock: June 14.
At 10 o'clock: June 22.

At 9 o'clock: July 7.
At 8½ o'clock: July 14.
At 8 o'clock: July 22.

At 9½ o'clock: June 30.

NIGHT SKY: JUNE AND JULY

some distance from the head, its cross motion, being at the same rate as that of the head when it left it, will be slower than that of a point on the line just mentioned, at an equal distance from the comet. The tail must, therefore, fall behind this line, and it is clear that the amount by which it lags will be greater, the farther we go from the head.

This may be made clearer by reference to the accompanying diagram. Suppose that the head of a comet, in its motion round the sun, is on three successive days at the points A, B, C. The tail particles given off on the first and second days would have

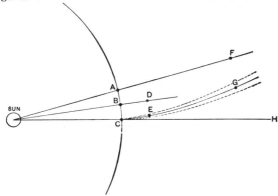

DIAGRAM SHOWING WHY A COMET'S TAIL IS CURVED.

been driven by the sun's repulsion from B to D in one day and from A to F (a much longer distance) in two days, if they had started without any motion of their own. But since they keep on moving laterally at the same rate as the head when they left it, the particles emitted on the second day will have moved from D to E, and those sent forth on the first day from F to G, by the time the third day arrives. The actual line of the comet's tail on that day will be the curve C, E, G, lagging far behind the line C H, which extends directly away from the sun.

The smaller the repulsive force, the more sharply curved the tail will be; as will be evident to anyone who will draw a second diagram, in which the comet is supposed to have moved twice as far in the time that a weaker repulsive force drives the particles over the same distances A F and B D.

As the repulsive force depends on the size of the fine particles composing the tail—which cannot be predicted—it is impossible to tell beforehand how much the tail will be curved. At this return Halley's comet was so placed (while it had any tail to speak of) that we saw the curve edgewise, and could not even estimate the sharpness of its curvature, until its delay in passing us revealed the facts. That there would be some delay was foreseen, but its amount was unexpectedly large, showing that the repulsive force on the tail of this particular comet is relatively feeble.

HALLEY'S COMETARY STUDIES.

HIS OWN ACCOUNT OF HIS INVESTIGATIONS ON ORBITS.

HALLEY's treatise, which bore the title "Astronomiæ Cometicæ Synopsis," (i. e. "A Synopsis of the Astronomy of Comets") was presented to the Royal Society in 1705, and was published in 1706 in volume 24 of the Society's Transactions, page 1882-1899. The following account of this work is taken from Volume IV. of Baddam's abridgment of the Memoirs of the Royal Society (London, 1739). The original spelling and punctuation are retained here.

The ancient Egyptians and Chaldeans (if we may credit Diodorus Siculus) being furnished with a long series of observations, could predict the rising, or appearing of comets; but since, they also are said by the same arts to have foretold earthquakes and tempests, it is past all doubt, that their knowledge in these matters was rather the result of astrological calculations than of astronomical theories of the motions of the celestial bodies; and the Greeks, who were the conquerors of both these nations, scarcely found any other sort of learning among them than this; so that it is to the Greeks themselves, as the inventors, especially to the great Hipparchus, that we owe this astronomy, which is now so greatly improved; but Aristotle's opinion (viz., that comets were nothing else than sublunary vapors or airy meteors) prevailed so far amongst the Greeks, that this sublimest part of astronomy lay altogether neglected; since none could think it worth while to observe, and give an account of the wandering and uncertain paths of vapours floating in the Æther; whence it is, that we have nothing certain handed down from the ancients concerning the motion of comets; but Seneca the philosopher, considering the Phænomena of two re-

markable comets of his time, made no scruple to place them amongst the celestial bodies, taking them to be stars of equal duration with the world itself; tho' he owns, that their motions were regulated by laws not then discovered; at length, he foretells (which has proved no vain prediction) that time and diligence would unfold these mysteries to some future ages, who would be surpris'd how the ancients could be so ignorant of them, after that some lucky interpreter of nature would have pointed out in what parts of the heavens comets wandered, and shewn what, and how great they were; yet almost all astronomers differed from Seneca in this; and Seneca himself has not left any account of the Phænomena of the motion, whereby he might support his hypothesis, nor assigned the time of their appearing, which might enable posterity to determine anything in this matter: So that after Mr. Halley had turned over several histories of comets, he could find nothing at all, that could give any assistance herein, before A. D. 1337, when Nicephorus Gregoras, an historian and astronomer of Constantinople, had pretty accurately described the path of a comet amongst the fixed stars; but he too loosely assigns the time, so that this undermined comet only deserved to be inserted in the catalogue, on account of its having appeared almost 400 years ago; the next comet A. D. 1472, which moved the swiftest of all, and came nearest to the earth was observed by Regiomontanus; this comet (so frightful on account both of the magnitude of its body and its tail) in the space of a day moved 40 deg. of a great circle in the heavens, and it is the very first of which, any proper observations have been handed down to us; for all those who considered comets before Tycho Brahe, that great restorer of astronomy, supposed them to be below the moon, and so took but little notice of them, imagining them to be no other than vapours: But in the year 1577 Tycho Brahe applying himself seriously to the study of astronomy, and having procured large instruments for making celestial mensurations, with greater exactness and certainty than the ancients could ever hope for; there appeared a pretty remarkable comet, to the observation of which Tycho vigorously applied himself, and he found by several unquestionable trials, that it had no sensible diurnal parallax; and consequently, that it was not only no aerial vapour, but much higher than the moon; nay, and might be reckoned amongst the planets for anything that appeared to the contrary, notwithstanding the cavilling opposition of some schoolmen; to Tycho succeeded the sagacious Kepler, who having the advantages of Tycho's observations, found out the true and physical system of the world, and vastly improved astronomy; for he demonstrated, that all planets revolved in planes passing through the centre of the sun, and describing elliptical curves, observing this law, that the area's of the elliptic sectors taken at the centre of the sun, in the Focus of the ellipsis, are always proportional to the times, wherein the corresponding arches are described; he also discovered, that the distances of the planets from the sun are in the sesquilateral ratio of their periodical times, or that the cubes of the distances are as the squares of the times; this great astronomer had the opportunity of observing two comets, one of which was very remarkable; and from his observations of these, he concluded, from several indications of an annual parallax, that comets move freely thro' planetary orbits, with a motion not much different from a rectilinear one, but which he could not determine: Next Hevelius, a noble emulator of Tycho, following Kepler's steps, embraced the same hypothesis of the rectilinear motion of comets, he himself having very accurately observed several of them; yet he complained, that his calculations did not altogether agree with the appearances in the heavens, and he was aware that the path of comets was incurvated toward the sun: At length, that extraordinary comet of 1680 descended from a vast distance, and as it were, in a perpendicular line toward the sun, and ascended from him again with an equal velocity; this comet appearing constantly for four months, by the peculiar and remarkable curvity of its orbit, seemed above all others the most adapted for investigating the theory of their motion; and the Royal observatories at Paris and Greenwich being founded some time before, and committed to the care of the most famous astronomers, the apparent motion of this comet was (as far as human sagacity could reach) very accurately observed by M. Cassini and Mr. Flamstead: Not long after, that incomparable geometrician Sir Isaac Newton, not only demonstrated that what Kepler had found, did necessarily obtain in the planetary system, but likewise that all the Phænomena of comets plainly follow from the same principles, which he fully illustrated by the above-mentioned comet of 1680; and at the same time shewed the way of geometrically constructing the orbits of comets, and to the surprise of all men solved a problem whose intricacy rendered it worthy of so great a genius; and he proves that this comet revolved round the sun in a parabolic orb in such a manner that the areas estimated at the centre of the sun were proportional to the times: Mr. Halley pursuing the steps of so great a man attempted (and he presumes not without success) to bring the same method to an arithmetical calculation; for having collected together all the observations of comets he obtained the following table,

THE ASTRONOMICAL ELEMENT OF THE MOTIONS, IN A PARABOLIC ORBIT, OF ALL THE COMETS HITHERTO OBTAINED.

Passage of Perihelion, London Time.	d.	h.	m.	Perihelion °	'	Asc. Node °	'	Inclination of Orbit °	'	Distance from Sun at Perihelion.
1337, June	2	6	25	37	59	84	21	32	11	0.40666
1472, February	28	22	23	45	34	281	46	5	20	0.54273
*1531, August	24	21	18	301	39	49	25	17	56	0.56700
1532, October	19	22	12	111	7	80	27	32	36	0.50910
1556, April	11	21	23	278	50	175	42	32	6	0.46390
1577, October	26	18	45	129	32	25	52	74	33	0.18342
1580, November	28	15	00	109	6	18	57	64	40	0.59628
1585, September	27	19	20	8	51	37	42	6	4	1.09358
1590, January	29	3	45	216	54	225	31	29	41	0.57661
1596, July	31	19	55	228	15	3.2	12	55	12	0.51293
*1607, October	16	3	50	302	16	50	21	17	2	0.58680
1618, October	29	12	23	2	14	76	1	37	34	0.37975
1652, November	2	15	40	28	19	88	10	79	28	0.84750
1661, January	16	23	41	115	59	82	30	32	36	0.44851
1664, November	24	11	52	130	41	81	14	21	18	1.02576
1665, April	14	5	16	71	54	228	2	76	5	0.10649
1672, February	20	8	37	47	0	297	30	83	22	0.69739
1677, April	25	00	38	137	37	236	49	79	3	0.28059
1680, December	8	00	6	262	40	272	2	60	56	0.00612
*1682, September	4	7	39	302	53	51	16	17	56	0.58328
1683, July	3	2	50	85	30	173	23	83	11	0.56020
1684, May	29	10	16	238	52	268	15	65	49	0.96015
1686, September	6	14	33	77	0	350	35	31	22	0.32500
1698, October	8	16	57	270	51	267	44	11	46	0.69129

Those marked with a star (*) are successive apparitions of Halley's Comet.

the result of almost immense calculation.

[Then follows a general table to compute the Motion of Comets in a parabolic Orbit, together with an explanation of its construction and use.]

It is to be observed that the five first comets, the third and fourth of which was seen by Petrus Apianus, and the fifth by Paulus Fabricus, as was the tenth by Mestlinus in the year 1596, have not the same degree of certainty with the rest, the observations not having been made with the proper instruments or the necessary exactness, and therefore, disagreeing with each other, they can by no means be reconciled with a regular calculation; Blanchini alone observed at Rome the comet Anno 1684; and the astronomers at Paris the last comet in 1698, whose path they have described in an unusual manner; this very obscure comet, tho' swift and pretty near the earth, escaped our observations: Mr. Halley forbore to insert into his catalogue the two remarkable comets that appeared, the one in November 1689, and the other in February 1702, for want of observations; for directing their course towards the southern part of the world, and being scarcely visible in Europe, they were not observed by persons equal to the task: It is to be observed that 11 of the comets in Mr. Halley's catalogue moved direct, i. e., according to the order of the signs; viz. those in the years 1532, 1556, 1580, 1585, 1618, 1652, 1661, 1672, 1680, 1684 and 1686; and that the other 13 were retrograde, i. e., moved contrary to the order of the signs.

Upon weighing all these things, and comparing the rest of the elements of the motions of these comets with each other, it will appear, that their orbits are disposed in no certain order; and that they are not confined like the planets to the zodiac, but that they move indifferently, every way both retrograde and direct; whence it is plain, that they are not moved by Vortices; the distances of this Perihelia are found to be sometimes greater and sometimes less; whence we have reason to suspect, that there are a great many more comets, which being at remote distances from the sun, and being obscure and without a tail, may for that reason escape our observation.

We have hitherto considered the orbits of comets as perfectly parabolic, from which supposition it would follow, that comets, being impelled by a centripetal force toward the sun, do descend from infinite distances, and by their fall acquire so great a velocity, as to convey them into the remotest spaces of the system, and by a perpetual Nisus tending upward, never afterward to return again to the sun; seeing then that the appearing of comets is very frequent, and that none of them is found to move in an hyperbola, or with a greater velocity than it would acquire in falling toward the sun, it is more credible, that they revolve about the sun in very excentric orbits, and return after very long periods; for, thus their number is definite and perhaps not so very great; and the spaces between the sun and the fixed stars are so immense, that there is room for a comet to perform its period, however large it may be: for the Latus rectum of an ellipsis is to the Latus rectum of a parabola, having the same Perihelian distance, as the Aphelian distance in the ellipsis is to its whole axis; but the velocities are in the sub-duplicate ratio of the same: wherefore, in very excentric orbits, this ratio approaches very nearly to a ratio of equality, and the small difference which arises on account of the greater velocity in a parabola, is very easily compensated in determining the situation of the orbit; therefore, the principal use of the elements of the motions in this table is, that whenever a new comet appears, we may by comparing the elements, know whether it is one of those that formerly appeared; and consequently, we may determine its period, and the axis of its orbit, and foretell its return; and Mr. Halley tells us that he had several reasons to induce him to believe, that the comet in 1531, which was observed by Apian, was the same with that described in 1607 by Kepler and Longomontanus, and which he himself had seen and observed upon its return again in 1682; all the elements agree, and there is no other difference than the inequality of their periods; which yet is not so considerable, as that it may not be ascribed to physical causes; for Saturn's motion is disturbed in such a manner by the other planets, especially Jupiter, that its periodical time is for some whole days uncertain; how much more may a comet be subject to such irregularities, whose orbit rises almost four times higher than Saturn's, and whose velocity, tho' never so little augmented, may change its orbit from an ellipsis to a parabola; that it was the same comet, is farther confirmed, from that observed, in the summer of 1456, to pass retrograde, almost in the same manner, between the sun and the earth; which tho' it was not observed astronomically by any, yet Mr. Halley con-

jectures, that was the same with the former, from its period and the manner of its transit; whence he ventures to foretell its return in 1758, and if this happens, there will be no further cause to doubt, but that the rest may likewise return; astronomers will therefore have a large field to exercise themselves in for several ages, before they can determine the number of so many and so great bodies that revolve round the common centre of the sun, and reduce their motions to certain rules: Mr. Halley was apt to believe, that the comet of 1532 was the same as that observed by Hevelius in the beginning of 1661; but Apian's observations, which are the only ones we have, are too inaccurate, to determine anything certain from them in so nice an affair: Sir Isaac Newton delivers a method of constructing the orbits of comets by three accurate observations, Philos. Natural. Princip. Mathemat. lib. III. which afterwards Dr. Gregory fully and clearly illustrated in the fifth book of his physical and geometrical astronomy.

Here one thing is to be observed, viz., that some of these comets have their nodes so near the annual orbit of the earth, that should it happen at the time of the return of a comet, that the earth was near its node, whilst the comet passes with an incredible velocity, it would also have a very sensible parallax, and which would be to the sun's parallax in a given ratio; whence, upon such like transits, there would be a very favorable opportunity (which yet seldom happens) of determining the distance of the sun from the earth; which hitherto could be concluded but very loosely, and that only by means of the parallax of Mars in opposition to the sun, or that of Venus in the Perigæum: and tho' indeed it is thrice greater than the parallax of the sun, yet it is scarcely perceptible with any instrument, and this use of comets was suggested by the famous geometrician Nio Facia; for the comet of the year 1472 had a parallax 20 times greater than that of the sun; and had the comet Anno 1618 arrived, about the middle of March at its descending node, or had the Comet Anno 1864 come a little sooner to its ascending node, being very near the earth they would have had still more sensible parallaxes; of all the comets there were none that approached nearer the earth than Anno 1680; for, upon a calculation, it was no further distant towards the north from the annual orbit, than the sun's semidiameter (or the radius of the moon's orbit, as Mr. Halley suppose) and that too in November 10th 1 hr. 6 min. P. M.; at which time it had been in conjunction with the earth as to longitude, there might have been observed in its motion a parallax equal to that of the moon: Mr. Halley leaves it to philosophers to discuss what consequences would arise from the appulse, contact or collision of the celestial bodies, which yet is not altogether impossible.

EDMUND HALLEY.*

THE MAN WHO DISPELLED COMETARY SUPERSTITIONS.

BY J. E. GORE, M. R. I. A.

In view of the return of Halley's comet, which has now been found by the telescope, some account of the great astronomer with whose name the comet is associated may prove of interest to the general reader.

The great French astronomer Lalande considered

* Knowledge and Scientific News.

Halley to be the greatest English astronomer of his time, and this opinion is certainly just. He was doubtless one of the greatest men of science that England, or, indeed, any country, has ever produced.

Edmund Halley was born in London on November 8th, 1656. From his earliest year, he applied himself with zeal to the study of mathematics and astronomy, and obtaining such instruments as his means permitted he studied the heavens, and took advantage of every fine night to improve his knowledge of astronomy.

Knowing that a considerable amount of excellent work had been done on the solar system and among the stars of the northern hemisphere by Cassini, Flamsteed, and Hevelius, Halley thought it better not to enter into competition with these great observers, and decided to turn his attention to the southern stars of which no good observations had then been made. It was reported that a Dutch observer named Houtman had observed these stars in the island of Sumatra, and that Blaeu, the globe maker, had used these observations in the correction of his celestial globes. But Halley, on examining these corrections, came to an unfavorable opinion about the observer and his instruments. Having consulted with some of his friends on the best station to choose for his proposed observations of the southern stars, Halley decided on St. Helena. His father, a citizen of London, promised to supply him with the money necessary for the purpose. He was recommended to King Charles II. by Williamson and Jones Moore, and the Indian Company, who had control of the island, promised him all the assistance he required.

Having decided on the undertaking, Halley had a sextant of 5½ feet constructed, somewhat similar to that of Flamsteed, and he procured telescopes and micrometers. The largest of the telescopes was 24 feet in length. He sailed for St. Helena in November, 1676, at the early age of 20, and after a satisfactory voyage of three months he arrived at the island.

Although he had heard good accounts of the climate, it proved very disappointing. Frequent rains and a constantly hazy sky hardly permitted any observations in the months of August and September, and although he lost no opportunity for observation he found that at the end of the first year he had only observed some 360 stars.

Notwithstanding these difficulties he consoled himself with the idea that his labors would not be wholly in vain, and he connected the stars he observed with others whose positions had been determined by Tycho Brahé. His catalogue of southern stars, reduced to the epoch of 1677, was published in London in 1679.

In forming his catalogue he had necessarily to identify the stars he observed, and in this connection he made the following remarks with reference to stars in the constellation Sagittarius: "It may appear strange that the stars which are before the left hock and the knees (in the ancient figure of a centaur), which Ptolemy called bright and of the second magnitude, are at present of the fourth at most. This appears to show if not corruptibility, at least the changeability of the celestial bodies. This diminution of light may be attributed to some accident. It is probable that for a similar reason the stars of the left leg and right knee are now so small that Tycho overlooked them." Halley further remarks that some stars which Ptolemy placed before (that is, west of) Piscis Australis, and of which four are noted of the third magnitude, are at present only of the fifth or sixth, and he asks if this is due to the diminution of their light in the lapse of time.

In addition to his work on the stars, Halley made some investigations on the moon's parallax, combining his observations at St. Helena with those made in Europe. He also made some researches on the theory of the moon's motion, which was not thoroughly understood in his time, and which were so necessary for the determination of longitudes.

On November 7th, 1677, Halley observed a transit of Mercury across the disk of the sun. From his observations, which were not, however, obtained under favorable conditions, he derived a value of the sun's parallax of 45 min., which is, of course, much too large, the real value being about 8.8 min. But transits of Mercury are not suitable for measures of solar parallax, and, indeed, Halley pointed out that transits of Venus are much better for the purpose.

In 1679, Halley paid a visit to Hevelius to satisfy himself as to the accuracy of observation which the latter had claimed for his method of pinnules without the aid of a telescope. Halley convinced himself that the errors of the observations made by Hevelius were less than had been supposed, and did not exceed a minute of arc.

In the year 1684 Halley paid a visit to his friend, Sir Isaac Newton, to confer with him about the law of gravitation which had been previously discovered by Newton. The result of the consultation between these great men led to the publication of Newton's famous work, the "Principia." The first book of this immortal work was presented to the Royal Society of London in April, 1686, and ordered for publication in May of that year. But as the finances of the society would not admit of paying for the printing of the work the expense was generously defrayed by Halley. The "Principia" was published in 1687, and it is pleasant to know that Halley's expenses connected with its publication were afterward refunded by the sale of copies of the work.

Halley wrote a work on comets. Among former writers on the subject he found that the first account of a comet of any use for his purpose was one by Nicephorus Gregoras, who in the year 1337 recorded carefully the apparent path of a comet in the sky. But he was not so accurate in his record of time, and it was only on account of its having dated back by over 300 years that Halley included it in his catalogue of comets. The great comet of 1472, which made a near approach to the earth, was observed by Regiomontanus. This comet described an arc of 40 deg. in one day, and is the first of which satisfactory observations were made. The comet of 1577 was well observed by Tycho Brahé. Kepler observed two comets, and from their parallax he concluded that comets pass freely through the planetary orbits along a path which differs but little from a straight line. These views were supported by Hevelius, but Borelli seems to have thought that comets moved in elliptical or parabolic orbits round the sun, and this theory was confirmed by the great comet of 1680, which the observations of Halley and Newton showed to be moving in a curved path round the sun.

In his researches on the motions of comets, of which he computed the orbits of 24, Halley noticed that those recorded by Appian in 1531, by Kepler in 1607, and by Halley himself in 1682, seemed to return after a period of seventy-five or seventy-six years. He thus suspected that these comets were one and the same. This suspicion seemed to be confirmed by the recorded appearance of bright comets in the years 1456, 1380, and 1305, the intervals again being seventy-five or seventy-six years. He rightly conjectured that the comparatively small differences in the recorded intervals were due to the disturbing action of the larger planets. This was especially so when he found that in the year 1681 the comet passed near the planet Jupiter, the attraction of which must have had a considerable influence on the comet's motion. Making due allowance for this disturbing influence of Jupiter, he computed that the comet would return to the sun's vicinity about the end of 1758 or beginning of 1759. Halley did not live to see his prediction fulfilled (he died in 1742), but the comet returned, and, according to the calculations of Clairaut, its nearest approach to the sun was due on April 13th, 1759, with a possible error of one month.

The comet actually passed through perihelion on March 13th, 1759, only one month earlier than the date fixed by Clairaut! This showed the accuracy of Halley's foresight, and the comet has ever since been known by his name, although he was not its original discoverer. The comet again duly returned in 1835, and its next passage through perihelion will take place about April 20th, 1910. This is now certain, as the comet was found on a photographic plate taken by Dr. Max Wolf on the morning of September 12th of the present year, within a few minutes of arc of the position predicted by the calculations of Messrs. Cowell and Crommelin of the Greenwich Observatory— a great triumph for astronomy and mathematics.

In addition to his merits as an astronomer, Halley was a good navigator. In 1698 he obtained command of a vessel and voyaged in the Atlantic Ocean for the purpose of making observations on the laws which govern magnetic variation. He returned to England in June, 1699. His lieutenant, who had given him much trouble, was dismissed, and Halley continued his voyage. He reached 52 deg. of southern latitude, when the ice compelled him to turn back, and after some adventures he reached London on September 7th, 1700. He had another voyage in 1701 and 1702, and then retired from traveling.

In November, 1703, Halley succeeded Wallis as Savilian professor of geometry at Oxford. In 1713 he was elected secretary of the Royal Society, and in 1721 he succeeded Flamsteed as Astronomer Royal at Greenwich. He was then 64 years of age, but still full of zeal and energy. When Halley took charge of the Greenwich Observatory he found it almost devoid of instruments, Flamsteed's executors having removed all those which belonged to him. He obtained a meridian circle, similar to that of Roemer, which he used for four years to observe difference of right ascension. In 1725 a quadrant of 5 feet radius was set up in the meridian, and his efforts were chiefly directed toward the improvement of the lunar tables for the determination of the longitude at sea.

Halley observed the great aurora, which occurred on March 16th, 1715, and his accurate prediction of the total eclipse of the sun on May 2nd, 1715, added much to his reputation.

He died on January 14th, 1742, at the good old age of 85.

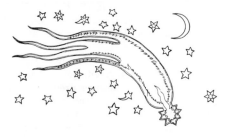

The Great Comet of 1577.

Comets

3

by Fred L. Whipple
July 1951

*Astronomers have recently constructed some plausible
theories about the composition and origin of these
periodic visitors to the center of the solar system*

THE "hairy" comet has long defied detailed scientific explanation and even today maintains some of its ancient mystery. Perhaps one of the reasons why comets remain so mysterious is that they have not received much general attention during the past 40 years. No really bright comet has displayed itself in Northern skies since the last appearance of Halley's in 1910. The constant discovery of faint new comets has, however, kept interest in the subject alive among a few astronomers, and progress has been made in explaining some of the puzzling features of the phenomenon. What are comets made of? Where do they come from? Why are they visible only when they are relatively close to the sun? I shall concentrate mainly on the attempts to answer these central questions.

The problem that has occasioned the most heated controversy is whether comets originate within the solar system or outside it. The problem arises from the eccentricity of the comets' orbits. Unlike the planets, which move about the sun in orderly and nearly circular orbits, all in nearly the same plane, comets have very erratic motions. They move in all possible directions, their paths forming a hodgepodge across the sky. And their orbits are extremely elongated instead of circular. The idea that comets might come from outside the solar system rested on the fact that some of their orbits are slightly hyperbolic, that is, open curves whose ends can never meet. But George van Biesbroeck of the Yerkes Observatory and E. Stromgren of Denmark and his collaborators have recently completed investigations which apparently settle the question. They traced back the paths of some 22 comets that

were observed to be moving in hyperbolic or practically parabolic orbits. They were able to prove that in every case these comets were traveling in elliptical orbits when they entered the central planetary region of the solar system. Those comets that acquired slightly hyperbolic orbits were thrown into such orbits by the gravitational attractions of the major planets, particularly Jupiter. Some of these comets, having gone off on a hyperbolic path, are now lost to the solar system forever, but we know that they *were* members of the solar system before the planetary perturbations disturbed their orbits.

These would not exclude the possibility that comets originally came from interstellar space and were captured by the planets and forced into elliptical orbits around the sun. But nearly 30 years ago the Princeton University astronomer Henry Norris Russell showed that this was unlikely to have taken place within the last 10 million years or so. He reasoned that if such captures had occurred recently and were still going on, we should observe many comets in hyperbolic orbits and should not observe such a high concentration of comets with almost parabolic orbits—comets having periods of millions of years. This conclusion has recently been verified in a more detailed research by J. J. van Woerkom at Leyden in the Netherlands.

Hence we may rule out the hypothesis that comets are now being captured or have recently been captured from a comet factory in interstellar space. They are truly members of the solar system, not interlopers. This conclusion immediately raises new questions. We know that comets must be relatively short-

lived; a number of them have disappeared even in the brief period that astronomers have been observing them. How, then, is the supply maintained? Have they been formed recently, or are they still being formed, in the solar system? If not, how could these frail phenomena have persisted through the three billion years of the solar system's lifetime?

THE answer seems to be that the solar system has an enormous population of comets extending very far into space. Nearly 20 years ago the Estonian astronomer E. Opik calculated that the sun's gravitational attraction could maintain a family of comets extending out as far as the nearest stars, some four light-years away, without the loss of a large fraction of these comets even during three billion years. To the objection that stars passing through this cloud of comets would rapidly destroy them he had an ingenious answer. A star passing through such a cloud is analogous to a bullet fired through a swarm of gnats. The bullet will eliminate only a small fraction of the gnats, without much disrupting the swarm. For comets the situation would be dangerous only if the passing star came close enough to the sun to attract the sun away from the comets. The likelihood of such a close approach is extremely small, even over such a long period of time.

Opik's important concept of a huge cometary swarm extending to the nearest stars escaped the notice of many astronomers until it was recently rediscovered by J. Oort of Leyden. Oort extended this line of reasoning to provide a clear picture not only of the manner by which the storehouse of comets is maintained

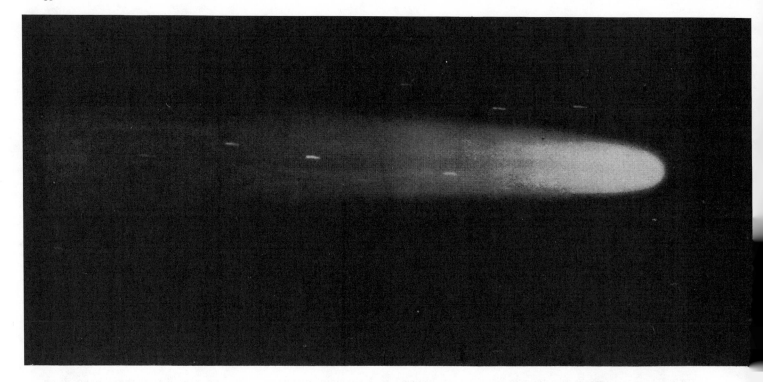

but also of the process by which withdrawals can be made. He called attention to the well-known fact that most comets disappear when they are twice the Earth's distance from the sun; very few have been visible as far away as Jupiter. Oort accepted the assumption that when comets are far from the sun they become completely inactive; that is, the particles and gases of which they are composed cease to vaporize or radiate energy. This inactivity in the "deep freeze" of space outside the planetary orbits is what enables comets to persist indefinitely in a state of hibernation.

Oort postulated that the cometary cloud may contain as many as 100 billion comets, very few of which come as close to the sun as the planets. Occasionally, however, the random passage of a star disturbs the motions of some comets sufficiently to make them swing into the sphere of gravitational attraction of Jupiter or another major planet. In this way comets are taken one by one from the "deep freeze" of the solar swarm and are pulled into relatively short-period orbits. Their hibernation period over, they become active and disintegrate into gas and meteoric particles during a few hundred or few thousand revolutions around the sun. The total supply of comets is so large, however, that despite these captures and losses into interstellar space, the comet cloud has persisted without serious depletion during the long lifetime of the solar system.

HAVING settled the problem of the storehouse and withdrawal system for comets, we may now turn to the problem of their physical nature. The most generally accepted theory is that comets are great flying "gravel banks"—masses of small solid particles held together loosely by gravity. As a comet moves around the sun the gravel bank is supposed to be slowly broken up by forces such as the sun's heat, radiation pressure, tidal disruption, rotation, and so on. This produces the streams of meteoritic material that are observed as meteor showers when the Earth crosses the orbits of certain comets ("Meteors," by Fletcher G. Watson; SCIENTIFIC AMERICAN, June 1951). As the comet approaches the sun, the sun's heat is assumed to drive gases from the surfaces of the particles, producing the gaseous spectra observed in comets.

This gravel-bank theory is rather effective in explaining many of the observed qualitative characteristics of comets. Not only does it explain meteor streams, cometary dust, cometary gases and increased activity near the sun, but it also explains how a comet may split in two, disintegrate and exhibit various observed irregularities in behavior. The theory is unsatisfactory, however, in most quantitative aspects. The postulated forces are generally too small to produce the observed disintegration, and it is difficult to see how the particles could carry enough gases, even if they replenished the supply by picking up gas molecules in interplanetary space, to account for the observed gaseous radiations.

These difficulties, along with other quantitative weaknesses in the theory, led me to search for a more adequate comet model. It occurred to me that a deep-freeze process might account for the formation of comets, just as it did for their storage in space. Why not assume that comets were made in the cold of

HALLEY'S COMET was photographed from the Chilean station of the Lick Observatory on May 11, 1910.

space from the solid particles floating there—the interstellar dust?

What solids would one expect to find in space? In the observable universe as a whole hydrogen is known to be by far the most abundant element, with helium second. Hydrogen and helium alone cannot freeze into solid particles, even in the cold of outer space, and helium does not combine with other elements. Hence we should expect to find that the solid particles in space are predominantly compounds of hydrogen with heavier elements, particularly carbon, nitrogen and oxygen, which are the next most abundant elements after hydrogen and helium. These compounds might be methane (CH_4), ammonia (NH_3) and water (H_2O). In the depths of space all three of these substances should be frozen solid as icy particles. I suggest that perhaps 70 to 80 per cent of the mass of a comet is composed of such ices. As the comet emerges from the deep freeze and approaches the sun, the ices vaporize. On the other hand, the remaining 20 to 30 per cent of the comet's mass, consisting of compounds of the heavier elements in space, do not vaporize appreciably even at rather high temperatures. These would be the particles that produce meteor showers.

Thus according to this model the nucleus of a comet would be a conglomerate of ices and solids—a dirty-looking ice including fine dust or even single molecules of a large number of compounds. In hundreds of millions of years the slight but continuous gravitational pressure within the nucleus would un-

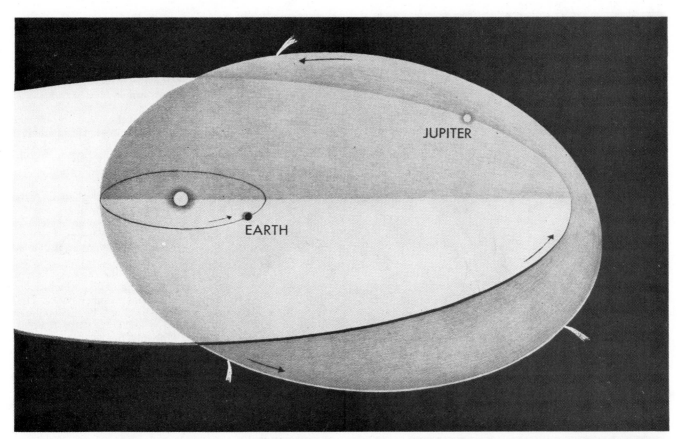

ORBIT OF A COMET is usually in a plane tilted with respect to that of the Earth and most of the other planets (*lighter disk*). Shown in this drawing is the orbit of Comet Giacobini-Zinner (*darker disk*), which intersects the orbit of the Earth and comes close to that of Jupiter. The comet's orbit is tilted at an angle of 31 degrees.

doubtedly consolidate the particles somewhat; this process may perhaps account for the sizable pieces of homogeneous material that occur in comets.

NOW at great distances from the sun, as we have suggested, the comet would be inactive. But as it approached the sun the solar heat would vaporize the material on the surface of the nucleus from the solid state. The escaping gases would carry meteoritic material with them to form a meteor stream along the tail and orbit of the comet. The gases themselves would then be acted upon by the radiation of the sun. Its ultraviolet light would break down the molecules of CH_4, NH_3 and H_2O into simpler forms. In this way we can account for the fact that the spectra of comets do not indicate the presence of CH_4, NH_3 or H_2O but do show the *radicals* of these compounds, such as CH, CH_2, NH, NH_2 and OH. Such radicals, which cannot be isolated in a terrestrial laboratory, probably are created by the rapid breakdown of the parent compounds by the ultraviolet sunlight.

The icy conglomerate model allows for the presence of certain other molecules that are found in comets. The carbon molecule C_2, so conspicuous in the tail of a comet, may well have been frozen directly in the comet's nucleus. Another conspicuous radical in comets is CN, often called cyanogen in analogy with the true cyanogen, C_2N_2. This radical probably arises not from C_2N_2 but from hydrocyanic acid, HCN. The theory could also explain the puzzling circumstance that the spectra of comets show the presence of free metals, including sodium, iron, nickel and chromium. These could either be frozen molecules imbedded in the conglomerate or products of the dissociation of compounds by ultraviolet sunlight.

Almost all comets show peculiar random fluctuations in brightness during their flight and a very rapid brightening as they come near the sun. Halley's and some of the other large comets erupt strong jets of gas. The model here proposed can easily explain this. We can assume that a porous crust of meteoritic material covers much of the surface of the icy nucleus, providing an insulating layer. As it approaches the intense heat of the sun, the heated nucleus tends to blow out gas and break holes through this crust or remove large sections of it. The underlying ices would then be exposed to direct solar heat and would produce a strong jet activity. The insulating layer may also account for the relatively long survival of comets after their capture by the inner solar system. By retarding the loss of their ices, the insulation may permit comets even as small as one mile or less in diameter to make hundreds or thousands of revolutions around the sun before they are completely dissipated.

A CERTAIN few comets show a most baffling departure from their timetables which astronomers have been at a loss to explain. An example is Encke's Comet, which revolves around the sun in three and one-third years, the shortest known period of a comet. In each revolution until about 1865 this comet arrived at its closest point to the sun about two and a half hours earlier than it was expected, and since then it has been, on the average, about one hour early. Such a deviation may seem too small to be taken seriously, but the astronomical observations are precise enough to assure that it cannot be due to errors in measurement. Another comet that behaved in similarly erratic fashion was Biela's Comet, which arrived late three times, mysteriously returned in two pieces in 1845 and thereafter was never seen again. It used to be supposed that the reason for the shortening of comets' periods was a resisting medium in space which slowed their motions and caused them to spiral in toward the sun in smaller and smaller orbits. But this notion had to be abandoned when it was discovered that some comets arrive behind schedule instead of ahead of time.

The icy conglomerate model suggests

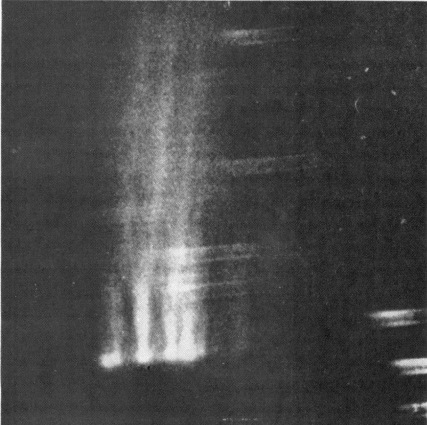

SPECTRUM OF A COMET tells something of its composition but little of its structure. These Yerkes Observatory photographs show Comet Morehouse in 1908. At the top is a direct photograph of the comet; at the bottom is the series of spectral images that are produced by photographing it through an objective prism. The spectra of comets show strong molecular radical lines.

a mechanism that may explain these strange phenomena. Besides revolving around the sun the nucleus of a comet may rotate on its own axis. Suppose that an erratic comet rotates on an axis that is not in the plane of its orbit. It continually turns a new face to the sun, as the Earth does. When the sun "rises" on a given face of the comet, it will take some time for the sun's heat to penetrate the insulating meteoritic layer and reach the underlying ices. It may be "afternoon" on the comet before the ices are heated sufficiently to vaporize and send out a stream of gas. At this late hour the outgoing vapor stream will be directed not toward the sun but at an angle from it. The stream of course has a jet-propulsion effect on the comet itself, giving it a push in a specific direction. Depending on the sense of the comet's rotation, this push will retard or accelerate the motion of the comet along its orbit. Thus the comet's period will be changed and it will arrive earlier or later than expected. Probably the jet propulsion of outgoing gases affects the motions of all comets to some extent. Halley's Comet, for example, returned in 1910 nearly three days behind schedule.

Calculations show that the loss of mass necessary to account for the jet-propulsion effects is very small; it would not materially reduce the lifetime of comets. Other quantitative checks on the icy-comet theory are possible. One of these concerns the zodiacal light, a sky glow which is conspicuous at the latitude of the northern U. S. during autumn evenings and contributes to the brightness of the solar corona seen at eclipses of the sun. The zodiacal light is produced by the scattering of sunlight by a cloud of small particles near the Earth's orbit. These particles must be slowly spiraling into the sun, because of a momentum effect of scattered sunlight known as the Poynting-Robertson effect. I find that about a ton of such particles should fall into the sun each second. This means that the zodiacal cloud must constantly be replenished by about this amount from some steady source of small particles. The comets, if the icy model is correct, release a total of some 30 tons of meteoritic material per second. Of this material only a part is in the form of particles of the necessary size and only a small fraction of them would survive the gravitational disturbances of Jupiter to contribute to the zodiacal cloud. Hence this check on the icy-comet theory seems satisfactory.

THE nucleus of a comet, to which the discussion has been confined so far, is not the principal part that one observes. In many comets the nucleus is too faint to be seen at all; only the hazy escaping envelope around it reflects or re-emits enough sunlight to be visible. The nuclei of the smaller comets are only

 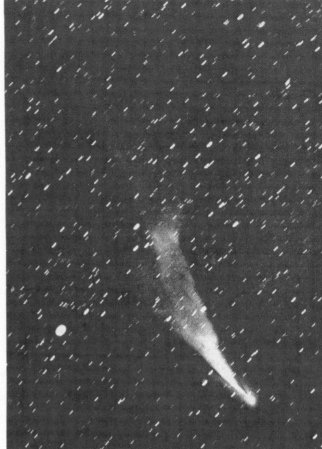

TAIL OF A COMET sometimes changes over a relatively short period. These Yerkes Observatory photographs show Comet Morehouse in 1908. The photograph at the left was made three hours before the one at the right.

a mile or so in diameter, but the visible head, or coma, may extend for thousands of miles beyond it. The nucleus of Halley's Comet is probably not over 10 to 20 miles in diameter; its tail, when last seen, was many millions of miles long.

The theory about the nucleus presented here does not help much to explain how the tails of comets are formed. It does, however, suggest a source for the large quantity of gas and exceedingly fine dust that constitute the tail. Much more theoretical and observational work is needed to explain with precision the processes in a comet's tail, particularly the high velocities of motion away from the sun. In general the pressure of sunlight on the fine dust and gas is clearly the basis of activity in the tail.

WE are still left quite in the dark as to the ultimate origin of comets. Where was the factory in which they were made located, and when did the sun acquire this magnificent assemblage of quite trivial bodies, whose combined total mass, in spite of their vast extent, is probably less than that of the Earth?

One possibility is that the solar system captured the comets from an interstellar cloud many millions of years ago, but we have no idea whether there are clouds in space with a concentration of material sufficient for the formation of comets. Another possibility is that the comets were formed along with the sun and planets from a great cloud of dust and gas that produced the solar system; this would follow naturally from the dust cloud hypothesis I once had the temerity to suggest (SCIENTIFIC AMERICAN, May, 1948). A third possibility is that they came from a disk of condensing gas about the sun which, according to one theory, gave rise to the planets. If so, the comets must have formed in the outer regions of the planetary cloud and may have been thrown into their helterskelter orbits, as G. P. Kuiper of the University of Chicago suggests, by the perturbations of the outermost planets.

Of one thing we can be certain: comets could not have been formed in the solar planetary system by the mechanism proposed by the British theoreticians R. A. Lyttleton and Fred Hoyle. Lyttleton suggests that whenever the solar system passes through an interstellar dust cloud, the attraction of the sun causes particles from the cloud to concentrate along the line of motion behind the sun. Such a concentration must certainly occur under ideal circumstances, but it is easy to show that planetary perturbations would appreciably disturb the motions of the particles. Thus their convergence upon this line would be so haphazard as to prevent the capture of appreciable aggregates of cometary matter by the solar system.

Be that as it may, we are at least somewhat in advance of Aristotle, who thought that comets dwelt in the upper atmosphere and gave off "hot exhalations" that dried the atmosphere and seared crops. Yet two thousand years ago Seneca, the Roman philosopher and historian, made a wise and prophetic comment about comets that could almost be written today. He said: "Why should we be surprised . . . that comets, so rare a sight in the universe, are not embraced under definite laws, or that their beginnings and ends are not known, seeing that their return is at long intervals? . . . The day will yet come when the progress of research through long ages will reveal to sight the mysteries of nature that are now concealed. The day will yet come when posterity will be amazed that we remained ignorant of things that will to them seem so plain."

Fred L. Whipple is professor of astronomy at Harvard University.

4

Review of
THE COMETS AND
THEIR ORIGINS
by R. A. Lyttleton

by James R. Newman
July 1953

*A general account of comets combined with an
interesting new theory of their origins*

THE COMETS AND THEIR ORIGINS, by
R. A. Lyttleton. Cambridge University Press ($5.00).

"OLD MEN and comets," wrote Dean Swift, "have been reverenced for the same reason; their long beards, and pretences to foretell future events." Anaxagoras and Democritus attributed comets to "the combined splendor of a concourse of planets." Aristotle, a renowned chronicler of old wives' tales about astronomy and meteorology, maintained that comets were exhalations from the Earth to the upper atmosphere. This hypothesis was so widely accepted that comets were not classified among the heavenly bodies in Ptolemy's *Almagest*.

Whatever the differences in the explanations of their physical nature, motions and causes, comets were until recent times universally regarded as presages, sometimes of happy augury, usually of death and disaster. The comet's sudden and mysterious appearance, its flaming flight across the sky, the swiftly changing aspect of its tail, its departure without a trace—all this inspired awe and fed superstition. When a comet is seen, "ther occurris haistily eftir it sum grit myscheif," said the *Complaynt of Scotlande* in 1549. Shakespeare (in Henry VI) wrote of "Comets importing change of Times and States." Milton's image in *Paradise Lost* is perhaps the most famous:

> On the other side,
> Incensed with indignation, Satan
> stood
> Unterrified; and like a comet
> burned,
> That fires the length of Ophiuchus
> huge
> In the arctic sky, and from his horrid hair
> Shakes pestilence and war.

The superstitions have all but vanished; myths about comets have been replaced by more fashionable irrationalisms. But the phenomenon itself is as impressive as ever, and as puzzling. The most recent edition of the *Encyclopaedia Britannica*, for example, reports that as late as 1946—when the article on comets was written—no plausible explanation of comet formation had been proposed. In *The Comets and their Origins* R. A. Lyttleton, a Cambridge mathematician, expounds a new, coherent and testable theory based on what he calls the "New Cosmology." Fred Hoyle, H. Bondi, and T. Gold have been in the forefront of the work underlying Lyttleton's study.

Lyttleton has explored the voluminous literature on comets and presents its main features in exemplary style. Tycho Brahe was the first to demonstrate what had already been conjectured by that strange combination of genius and muttonhead, Jerome Cardan: that comets are true celestial objects, far more distant than the moon. Tycho suggested that the path of the daylight comet of 1577 was a circle. Johannes Kepler, not realizing that comets could return—and were therefore subject to the rules of planetary motion which he himself had so brilliantly unfolded—concluded that they moved in straight lines. The German astronomer Johannes Hevelius conjectured their path as parabolic. It was Edmund Halley who, with the aid of Newton's theory of gravitation, finally solved the problem of cometary orbits. And the incomparable Robert Hooke, who rarely, if ever, pronounced on scientific subjects without saying something sensible, suggested that comet tails were formed by the pressure of the sun's rays. He was extraordinarily prescient: this is the view now generally adopted by astronomers.

Comets move in highly complicated three-dimensional curves. When these are simplified for computational purposes into so-called two-dimensional "osculating" orbits, it is found that the orbits are conic sections: hyperbolas, parabolas, ellipses. The osculating orbit is the path a comet would follow if it were subject only to the simple inverse-square attraction of the sun. While the sun is the dominating influence and lies at the focus of the conic, planets such as Jupiter cast the hem of their gravitational mantle over the comet, thus producing perturbations in its motion. In some instances the effect is very severe and results in a drastic change of orbit. Most osculatory orbits are parabolas, but a slight decrease or increase in the comet's speed converts the orbit into an ellipse or a hyperbola, respectively.

Astronomers have now determined the orbits of about 1,000 comets. The task is delicate and difficult. When a comet first comes into view, a provisional path is calculated on the basis of its behavior for a few days. It is then kept under constant surveillance so that its path with respect to the Earth may be computed with increasing precision. A crucial reference datum in this computation is the comet's perihelion point: it is necessary to ascertain the exact time at which the comet passes or will eventually pass in closest proximity to the sun. For a comet to be periodic, its orbit obviously must be elliptical; on a hyperbolic or parabolic journey the traveler will come to us literally out of the nowhere and vanish forever into the beyond. Yet it is precisely this determination—dependent on minute differences of curvature and subject to various circumstances that limit the accuracy of observation—which is so difficult to settle. Careful tracking may lead to the conclusion that a cometary orbit is hyperbolic near the sun; however, when the orbit is "extended further outwards from the observable part and backwards in time, by calculations making due allowance for the influence of the planets (Jupiter is usually the main perturbing agency), in every case it has been found that the comet has in fact come in from a finite distance, and is therefore to be regarded as a reappearance of a permanent member of the solar system as far as its orbital motion is concerned." Since Kepler's third law—that the square of the time of revolution of a planet is proportional to the cube of its mean distance from the sun—describes the time of revolution of a comet in its orbit, it is easy to see that the period of a comet calculated from a necessarily limited arc is "only weakly determined." This uncertainty helps to explain the extraordinary computational differences among leading astronomers. To cite only one illustration: the comet of 1680 has, according to the most accurate reckoning of the German astronomer Johann Franz Encke, a period of 8,814 years—not 170 years, as found by Leonhard Euler, or 575 years, as Halley calculated, or 5,864 years, as the Frenchman Alexandre Pingré said.

It must not be supposed that more than a small fraction of the comets chasing around the solar system ever have or ever will be seen. Even short-period comets (those traveling their circuit in less than 100 years) may disappoint us after a number of visits and fail to return. Brorsen's comet of 1846 (period 5.5 years) was not seen again after 1879, and Holmes's comet of 1892 (period 7 years) was not found in 1919 or 1928. Others, such as Encke's comet in 1944, are "missed through unfavorable circumstances but rediscovered at a later return." Halley's comet (period about 77 years) has been reasonably punctual for at least six centuries and may be looked for again in the spring of 1986. Comets of moderate period—of which about 40 are known—pay their homage to the sun every 100 to 1,000 years. But the great majority of comets have long orbital periods averaging about 40,000 years, according to the noted British expert A. C. D. Crommelin. It has been estimated that at least 300 long-period comets come to perihelion each century; if, then, the 40,000-year average is adopted, "we arrive at the amazing but inescapable conclusion that there must be at least 100,000 comets in the solar system with perihelion distances sufficiently small for them to become eventually observable." Moreover, there are many more with perihelion distances too great for the comets to be seen with present equipment if they remain in their existing paths.

About six or seven comets are discovered each year. The way to find one is to have the qualities which made a success of Phil the Fiddler—industry,

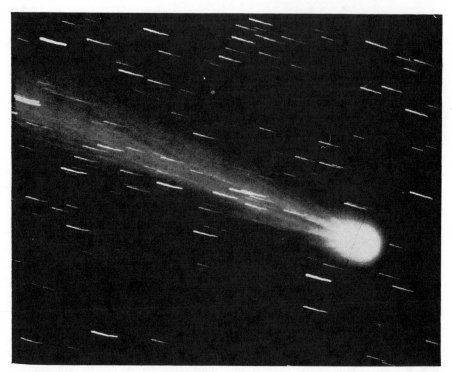

Daniel's Comet of 1907 had a fan-shaped tail with several "rays"

zeal and attentiveness. Good equipment and luck also help. In 1896 the U. S. astronomer Charles Perrine was at the Lick Observatory making loving observations of a comet he himself had discovered when he received a telegram from Kiel stating the position of the comet at that moment. The telegram had been jumbled in transmission and gave an entirely wrong position, more than two degrees from the correct one. Perrine, not knowing the message had been twisted, pointed his telescope to the indicated place—and found a new comet. The devotion and perseverance of astronomers is typified by the almost incredible labors of Joseph de Lalande and his staff in computing the date of the return of Halley's Comet in 1759. Monsieur Lepaute, one of Lalande's assistants, tells the story:

"During six months we calculated from morning till night, sometimes even at meals; the consequence of which was that I contracted an illness which changed my constitution during the remainder of my life. The assistance rendered by Madame Lepaute was such that without her we never should have dared to undertake the enormous labor with which it was necessary to calculate the distance of each of the two planets, Jupiter and Saturn, from the comet, separately for every degree, for 150 years."

The prediction was that the comet, having been delayed 100 days by the influence of Saturn and 518 days by that of Jupiter, would arrive at perihelion April 13, 1759. The actual date was only 32 days earlier, a tribute no less to the skill of the computers than to the theory.

What is a comet? Henry Norris Russell describes it as a "loose swarm of separate particles" accompanied by dust and gas. An observer sees it as a queer object with a head that is no head (and which, in any case, is sometimes missing), and a tail that conforms to no definition of a tail found in any dictionary (and may also be missing). The head, when it is present, consists of a faintly luminous cloud, called the coma, enveloping a bright something called the nucleus. The coma is transparent and undergoes fantastic deformations as it passes the sun; the nucleus, thought to be made of "some kind of changing concentration of small particles," also can be expected to transform itself like a jinni—a fact which astronomers conveniently explain by classifying the nucleus as no more than an "apparent phenomenon." The most spectacular feature of a bright comet is its tail. This appendage has been known to extend more than 200 million miles and nearly 180 degrees across the sky. Comets, like young women presented at the Court of St. James's, seem to put on their tails as they approach the sun. As a general rule, the closer the perihelion distance, the more impressive the tail. As a comet approaches the sun the tail streams behind, but beyond perihelion it precedes the comet. This curious behavior is explained on the theory that the pressure of the sun's radiation affects the particles of the tail in the way the wind

Halley's Comet of 1066 was depicted in the Bayeux tapestry

affects a plume of smoke. A comet can have more than one tail: Borelli's (1903) is said to have shown nine.

Comets are luminous partly because the small solid particles of which they are composed reflect, diffract and scatter the sun's light. But their spectrum, besides showing the familiar solar Fraunhofer lines, exhibits bright bands due to the emission by molecules of energy originally absorbed from solar radiation.

There is considerable variation in the sizes and shapes of comets and in their masses. A few cometary giants are greater in volume than the sun itself. The more normal specimens range in diameter from 20,000 to 200,000 miles. In 1909 Halley's comet, which is not atypical, had an observed coma 14,000 miles across when it was three times the Earth's distance from the sun; at a distance of two astronomical units the coma had swelled to 220,000 miles; at perihelion it had shrunk to 120,000 miles; later, at one unit distance, it had increased again to 320,000 miles. The prime example of shape-changing capacity is Biela's comet, a short-period wanderer (6.6 years). On its visit in 1846 it first startled observers by making its entrance in pear-shaped form; 10 days later it shattered the self-confidence of astronomers by dividing into two separate comets which "continued to travel in practically the same orbit, one preceding the other by about 175,000 miles." The twins appeared again in 1852, now eight times farther apart. They have not been seen since.

Biela's twinning, by the way, was discovered in the same year as the planet Neptune. Both events provided opportunities for the hapless Cambridge astronomer James Challis to demonstrate his exquisite incompetence. Challis managed to miss finding the planet Neptune

after John Couch Adams had told him where and when to look for it. His excuse was that he was searching for comets. To complete the tableau, he observed the twinning of Biela's comet but attributed it to an optical illusion and failed to publish his finding, explaining his oversight later on the ground that he was too busy looking for Neptune.

I think it was Sir John Herschel who said that a comet could easily be packed into a portmanteau. Certainly the mass of a comet is insignificant compared to that of other celestial objects. The average comet, Lyttleton suggests, is only one 10-billionth as massive as the Earth. Its weight comes to about a million million tons—too large for a portmanteau, but too small to produce observable gravitational perturbations. Spread this mass of small stones or rocks through a volume equal to that of the sun and there will be plenty of empty space between adjacent stones. On this assumption it is not surprising that comets are transparent. Lyttleton puts the density of the tail at "perhaps far less than" a trillionth of a trillionth of a gram per cubic centimeter, and supposes it to be made of a mixture of dust and gas. Meteors are related to comets; one plausible hypothesis is that they are simply the "debris of disintegrated comets."

The Lyttleton group's new theory of how a comet originates runs something like this: Genesis begins when the sun passes through a galactic dust cloud formed of material ejected in the explosion of a supernova. The sun's gravitational attraction starts these particles flying in orbits which are hyperbolic relative to the sun. Fix your attention now on an imaginary line through the center of the sun parallel to the direction of its motion—the so-called accretion axis. The particles of the cloud will converge toward this axis and will collide

when they are at and near the axis at points behind the sun. If we consider two inelastic particles of the same mass moving from points on directly opposite sides of the accretion axis, and "symmetrically placed with respect to the direction of motion of the sun," it can be shown not only that a head-on collision will occur on the axis, but that the results of the impact will be to nullify the transverse components of the particles' motion, to leave unaffected the radial components away from the sun along the axis, and "to reduce the originally hyperbolic energies of the particles to elliptic values, and thus bring about their capture by the sun." The multiplication of this and similar effects, involving hordes of colliding particles, creates a stream of material trailing behind the sun along the path of its motion. Part of the stream, however, flows toward the sun and part away from it, because out to a certain distance the sun's gravitational pull draws the colliding particles inward along the accretion axis, while beyond a certain point material arriving at the axis has a radial velocity sufficient to overcome the sun's captive power. These particles will escape.

According to the Lyttleton theory the stream flowing toward the sun is acted upon by two opposing forces: the sun's gravitational field and the internal gravitation of the material within the dense stream itself. The latter pulls the stream together lengthwise. We may assume, says Lyttleton, that irregularities of density in the dust cloud, and the general unstable nature of the accretion process, produce in the stream centers of attraction around which particles tend to cluster. While the internal gravitation is of negligible importance where the stream is close to the sun, farther out it is sufficient to promote not only the formation of local clusters but a lumping together and separation of the stream into segments. It is these segments which, according to the theory, develop into comets. One asks why, since this entire part of the stream is flowing toward the sun, the comets are not sooner or later swallowed up by it. Lyttleton answers that much of the stream's material does fall into the sun, but some of the comets escape this fate by the attraction of planets, particularly Jupiter and Saturn. The effect of their gravitational fields is to endow nascent comets which are favorably placed with a sufficient angular momentum to sweep clear of the sun. He estimates that even if only a small percentage of all comets forming in the stream avoid extinction, "an average cloud might easily produce several thousand comets" that will survive.

It is impossible not to admire the working out of Lyttleton's system. It is coherent and—for the average reader, at any rate—excitingly persuasive. Astronomers and cosmologists, even those favorably disposed to the accretion hypothesis, will be harder to persuade; some, I am sure, will look at Lyttleton's conclusions with a fishy eye. None, however, will overlook the important fact that the central idea involves hypotheses "that can be subjected to quantitative tests—a feature hitherto completely absent from cometary theories." While these tests can yield confirmation only within orders of magnitude, their results are not to be despised, for as the British physicist Harold Jeffreys once pointed out, most incorrect physical hypotheses fail such tests "by many powers of 10."

This book is not always easy to read, but I recommend it highly. It impressed me so much that I now appreciate the feelings of a certain enthusiastic young lady from New Jersey (reported by Mary Proctor in her book on comets) who on the appearance of Halley's comet in 1910 declared her intention of following it "wheresoever it went." Only "temporary seclusion" in an asylum deterred her from this gallant pursuit.

The Tails of Comets

5

5

5

by Ludwig F. Biermann and Rhea Lüst
October 1958

*It has long been assumed that the gases of a comet
tail are pushed away from the comet by the pressure
of light from the sun. It now appears that many
tails are caused by a wind of charged particles*

The planets move across the sky with stately regularity, but new comets appear and disappear unpredictably. It is no wonder that comets have traditionally been surrounded by an atmosphere of mystery. The mystery has been enhanced by their luminous tails, which earlier peoples took as omens of war and pestilence. Even today comets and their tails have not been fully explained, but the labors of astronomers have clarified many of their puzzling features.

In 1951 Fred L. Whipple of the Harvard College Observatory surveyed for the readers of this magazine our knowledge of comets at that time [see "Comets," by Fred L. Whipple; see page 29 of this reader]. The present article will be primarily concerned with recent studies of the tails of comets. At the beginning, however, it may be well to review what we know of comets in general.

It appears that all comets are members of the solar system; none are interlopers from outer space. According to a theory developed by the Dutch astronomers Jan H. Oort and J. J. van Woerkom, there is an enormous number of comets—probably 100 billion. They form a vast cloud at a distance of 50,000 to 100,000 astronomical units from the sun—almost as far as the nearest stars. (One astronomical unit is the mean distance between the sun and the earth.) A comet in the cloud moves around the sun in a huge orbit, one circuit of which takes millions of years. Occasionally, however, the gravitational attraction of a passing star may disturb the orbit of a comet so that the comet comes closer to the sun. Then the pull of the larger planets, notably Jupiter and Saturn, may further distort the orbit so that the comet makes a complete turn around

the sun in as little as a few years. Sometimes the orbit may be changed to an open hyperbola, and the comet escapes from the solar system entirely.

The comet itself is normally a conglomeration of solids. Held together by their mutual gravitational attraction, they form a nucleus perhaps a mile to 50 miles in diameter. According to Whipple's picture, the nucleus consists mainly of ices of water (H_2O), methane (CH_4) and ammonia (NH_3). Interspersed among these frozen compounds of the lighter elements are molecules and particles of heavier elements. When the comet is far away from the sun, its ices are kept in a "deep freeze"; when it approaches the sun, they begin to vaporize. The escaping gases surround the nucleus with an envelope (called the head or coma) from 10,000 to more than 100,000 miles in diameter. They may ultimately form the comet's tail, which in some cases extends 100 million miles.

Often the comet does not develop a tail at all. Whether or not the tail forms depends mainly on how close the comet comes to the sun; obviously more material will evaporate from the comet when it is near the sun than when it is farther away. But the formation of a tail also depends on two other factors: the properties of the individual comet (*e.g.*, its chemical composition) and the changing activity of the sun.

Comet tails vary greatly in appearance from comet to comet, and the tail of an individual comet may change from time to time. The tails have been classified into three main groups. Tails of Type I are long and straight; within them there are often threadlike streamers, knots and other structures. Spectra of such tails indicate that they consist

mainly of ionized molecules of carbon monoxide (CO^+), nitrogen (N_2^+), carbon dioxide (CO_2^+) and the hydrocarbon radical CH^+. Of these molecules those of ionized carbon monoxide appear to be by far the most abundant. Tails of Type II and Type III are more or less curved; most of them are shorter than tails of Type I. They are fuzzy and have little or no internal structure. Their spectra show no lines of ionized molecules; they are probably composed largely of gases which are not ionized and of dust. Often both kinds of tail—the straight and the curved—appear simultaneously in one comet [*see photographs on next two pages*].

What forces act on comets to produce the varied patterns of their tails? All comet tails point predominantly away from the sun, so it would seem reasonable to look for the origin of these forces in the sun itself. Of course the sun exerts a strong gravitational pull on all the matter in a comet. But there must also be a repulsive force which pushes the matter in the tails away from the sun.

In order to get a clearer picture of how the repulsive force operates, astronomers have measured the acceleration of matter in comet tails. This is done by noting on a photographic plate the position of an individual structure in a tail, and then observing on plates made at successive intervals how far the structure has moved. It turns out that the acceleration varies with the kind of tail. The acceleration of matter in a tail of Type I requires a repulsive force of the order of 200 times (occasionally up to 2,000 times) greater than the attractive force of the sun's gravitation. The acceleration in a tail of Type II requires a force equal to or perhaps twice as strong as the sun's gravitation; the acceleration

COMET MRKOS was photographed on August 23, 1957, with the 48-inch Schmidt telescope on Palomar Mountain. In this negative print the light areas are dark, and *vice versa*. Such prints are used by astronomers to increase contrast and accentuate faint objects.

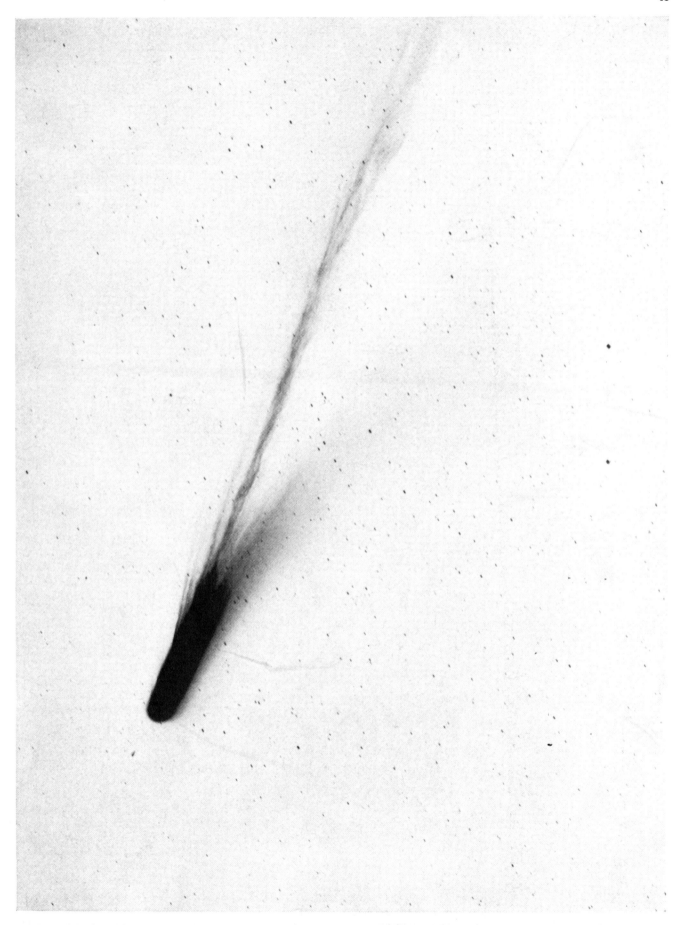

SAME COMET was photographed with the 48-inch Schmidt telescope two days later, when the structure of its tail had changed considerably. The long, thin streamers comprise a tail of Type I. The faint, curved dark area to right of this tail is a tail of Type II.

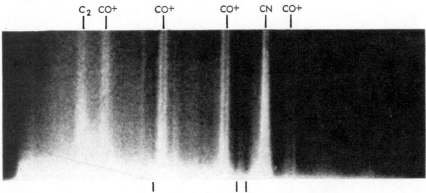

THREE TYPES OF COMET TAIL are depicted in these schematic drawings. Type I tails are long and straight; Type II tails are curved; Type III tails are also curved but shorter.

C₂ CO⁺   CO⁺   CO⁺   CN CO⁺

ABSORPTION LINES OF SOLAR SPECTRUM (SCATTERED SUNLIGHT)

SPECTRUM OF COMET MRKOS was made on August 28, 1957, by George H. Herbig of the Lick Observatory. The bright horizontal area at the bottom of the photograph is the spectrum of the nucleus of the comet. The vertical streaks are bright and dark lines in the spectrum of the comet's tail. The bright lines reveal the presence in tail of carbon (C_2), positively charged ions of carbon monoxide (CO^+) and the radical of carbon and nitrogen CN.

in a tail of Type HI, a force even smaller than the sun's gravitation.

Now astronomers have long assumed that the repulsive force was the pressure of light from the sun. Theoretical studies have indicated, however, that the force exerted by light on molecules or dust particles in the tail of a comet would be at most a few times greater than the gravitational force of the sun. Thus light pressure may account for the tails of Type II and Type III, but not for those of Type I. To explain the large accelerations of matter in the spectacular tails of Type I another force is needed.

Is it perhaps the ultraviolet radiation of the sun, as distinct from the sun's visible light? Sunspots are associated with an increase in the sun's ultraviolet radiation, and the German astrophysicist Max Beyer found indications that they are also associated with an increase in the brightness of the heads of comets. It is possible that this energetic radiation enhances the photochemical reactions which give rise to the molecules and ions of the tails. But a closer examination shows that ultraviolet radiation is even less able than visible light to account for the large accelerations of Type I tails.

What about the particles of matter which are ejected by the sun? In recent years much has been learned about this corpuscular radiation [see "Corpuscles from the Sun," by Walter Orr Roberts; SCIENTIFIC AMERICAN, February, 1955]. We know that it consists of electrons and positive ions in equal numbers—what the physicist calls a "plasma." These oppositely charged particles must surely interact with the positive ions of the long Type I tails. Can they account for the acceleration of matter in these tails?

The corpuscular radiation of the sun originates in the vicinity of a group of sunspots, but often it persists much longer than the spots. Sometimes, however, a brilliant eruption in the lower atmosphere of the sun gives rise to an unusually strong blast of corpuscles. When the corpuscles reach the earth, they cause a "storm" in the earth's magnetic field. By correlating the time of the eruption with the beginning of the magnetic storm, we can calculate the velocity of the corpuscles: they move at speeds up to 1,000 miles per second. The German astrophysicist Albrecht Unsöld and the British geophysicist Sydney Chapman have estimated the density of the corpuscles in these powerful streams: in violent magnetic storms it goes as

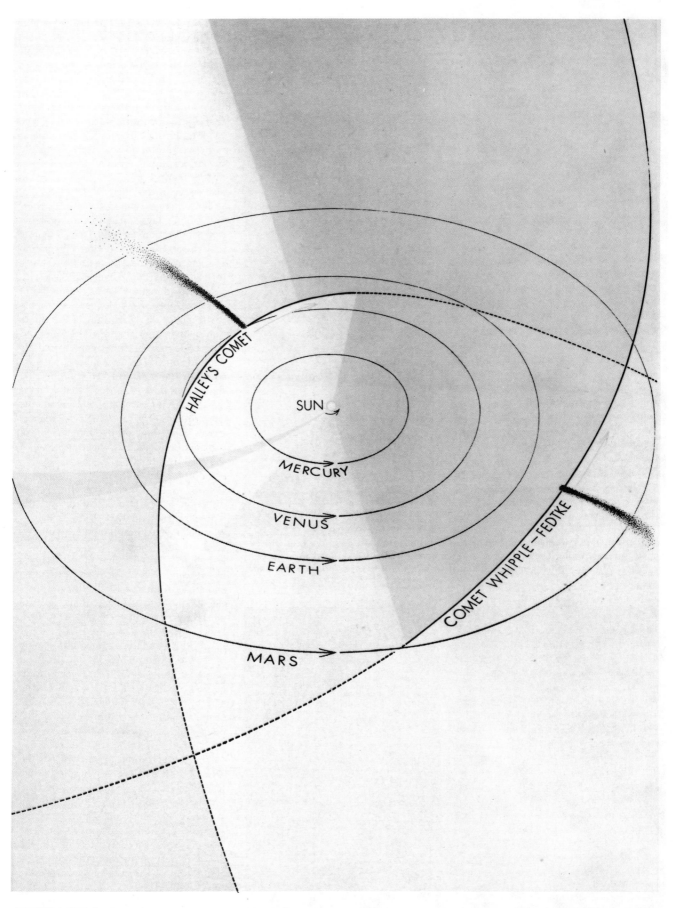

ORBITS OF TWO COMETS are depicted with respect to the orbits of the four inner planets. The planets move in the plane of the ecliptic. The sun also turns on its axis approximately in this plane. The orbits of the comets are inclined to the plane of the ecliptic. The curved hatched area represents a stream of particles emitted by the sun. The tails of the comets are depicted schematically.

HALLEY'S COMET was photographed on May 6, 1910, from a station of the Lick Observatory in Santiago, Chile. The short streaks which appear in the background were made by the images of stars, which remained stationary while the telescope followed the comet.

COMET WHIPPLE-FEDTKE was photographed in 1942 from the Sonneberg Observatory in Germany. The large spot above the tail of the comet is the image of a bright star.

high as 100,000 corpuscles per cubic centimeter.

Sometimes storms in the magnetic field of the earth occur at intervals of 27 days. Since it takes 27 days for the sun (as it is seen from the earth) to make one turn on its axis, it is assumed that these recurrent storms are caused by long-lasting streams of corpuscular radiation. If there is a relationship between corpuscular radiation and comet tails of Type I, periodic changes should also occur in the tails.

The changes will not come at intervals of 27 days, because the sun's period of rotation as seen from a comet will differ from the period of rotation as seen from the earth. Actually the sun turns on its axis once every 25 days with respect to the stars. To us the period of rotation appears to be longer because the earth moves along its orbit in the direction of the sun's rotation. The length of the period as seen from a comet similarly depends on the comet's motion with respect to the sun. Further complicating the situation is the fact that comets usually do not move in the plane of the ecliptic—the plane in which the planets move along their orbits, and which approximately coincides with the plane of the sun's rotation.

Consider the following two examples. We regard a comet as moving in the direction of the sun's rotation and the planets' orbital motion. Viewed in this way, Comet Whipple-Fedtke (discovered in 1942) moves in an orbit which is inclined to the plane of the ecliptic by an angle of 20 degrees. From the comet, as from the earth, the sun's period of rotation is longer than the true period. Viewed in the same way, Halley's Comet (last seen in 1910) moves in an orbit which is inclined at an angle of 162 degrees. Thus for all practical purposes Halley's Comet moves in a direction opposite the direction of the sun's rotation, and is inclined at an angle of only 18 degrees. From the comet the period of the sun's rotation is several days *shorter* than the true period.

The acceleration of matter in the Type I tail of Comet Whipple-Fedtke was measured at various times; the largest accelerations were observed on March 3, 1943, and on March 29 of the same year. Photographs made on these dates also show unusual features in the tail. The dates coincide with the period of rotation of the sun as seen from the comet: 26½ days.

On April 22, 1910, the acceleration of matter in the Type I tail of Halley's Comet was 240 times greater than the acceleration of matter falling toward the sun; between June 5 and June 9 it was of the order of 1,000 times greater. During all other observations the acceleration was considerably smaller. The interval between April 22 and June 7 corresponds exactly to two turns of the sun as seen from the comet. The intermediate date in May fell in an interval when the comet was so close to the sun that it could not be satisfactorily observed.

If corpuscular streams caused these cometary events, we would expect to find that the events were correlated with magnetic storms on the earth. Both comets were near the plane of the ecliptic, and the same corpuscular streams which hit them should also have played over the earth. On March 29, 1943, when large accelerations were observed in the tail of Comet Whipple-Fedtke, there was indeed a major magnetic storm; presumably the stream first hit the earth and some hours later the comet. The data of 1910 are not so clear-cut, because of a generally higher level of solar activity at that time, but there does appear to be a statistical correlation between the accelerations in Halley's Comet and magnetic disturbances on the earth.

This evidence (and some we have not discussed here) would seem to show that the sun's corpuscular radiation does have an effect on the tails of comets. But exactly how does it exert this effect?

The simplest mechanism would be ordinary friction. Here one would have three gaseous fluids moving through one another: the solar ions (mostly nuclei of hydrogen), the cometary ions (mostly molecules with one electron removed) and the free electrons of both the solar plasma and the cometary plasma. Calculations have shown that, if the temperature of the electrons is 10,000 degrees absolute (10,000 degrees centigrade above absolute zero), a solar plasma of 100 billion corpuscles per square centimeter per second would give the cometary ions a frictional acceleration of one meter per second per second.

The coupling of the solar corpuscles and the cometary ions may be enhanced by magnetic fields carried along with the corpuscles. There is evidence for such fields in the long, thin streamers of Type I tails. These streamers are sometimes as much as 500,000 miles long and only 500 miles wide. Even if one assumes that the temperature of the ions in a streamer is as low as 300 degrees absolute, their random thermal motions should make the streamer much wider. It is possible that the charged cometary molecules are held in narrow bundles by the magnetic field of a corpuscular stream. Curious helical structures in tails of Type I also suggest the presence of a magnetic field. We are presently studying photographs of Comet Mrkos, which came into view last year, to gain some insight into these relationships.

COMET AREND-ROLAND was photographed on April 27, 1957, with the 48-inch Schmidt telescope on Palomar Mountain. The projection to the left of the comet is not material projected toward the sun but is cometary debris in the plane of the comet's orbit.

The Nature of Comets

by Fred L. Whipple
February 1974

Although Comet Kohoutek was a disappointment to visual observers, it provided a good opportunity for the study of one of the objects that may be relics of the cloud from which the sun and the planets formed

Comet Kohoutek, like other comets, is a celestial fountain spouting from a large dirty snowball floating through space. The fountain is activated and illuminated by the sun. It is greatly enhanced because it is spouting in a vacuum and essentially in the absence of gravity. We see the fountain as the comet's head and tail. The tail can extend for tens of millions of miles, but we never see the snowball, whose diameter is only a few miles.

The word "comet" comes from the Greek *aster kometes*, meaning long-haired star. The tail of the comet is of course the hair; the head, or coma, of the comet could be considered the star. Within the coma is the snowball: an icy nucleus that moves in a huge orbit under the gravitational control of the sun. The nucleus spends almost all its lifetime at great distances from the sun, hibernating in the deep freeze of space. When its orbit swings it in toward the sun, its surface begins to sublime, or evaporate, and the sublimated gas flows into space. Pushing against the weak gravity of the relatively small nucleus, the outflowing molecules and atoms carry with them solid particles. Thus does the nucleus give rise to the gaseous and dusty cloud of the coma.

The sun floodlights the dust and gas of the coma, making the comet visible. Some comets are very dusty. Most of their observed light is simply sunlight scattered by the dust and is slightly reddish. Other comets contain little dust. Since molecules and atoms in a gas scatter light feebly, such gaseous comets become bright only through a double process. First the ultraviolet radiation from the sun tears the molecules apart; water, for example, is dissociated into hydrogen (H) and the hydroxyl radical (OH). Then the atom or the broken molecule can fluoresce, that is, absorb solar light at one wavelength and reradiate it at the same wavelength or (more usually) at a series of longer wavelengths.

Almost all the light from gaseous comets comes from such bands of wavelengths, which are mostly emitted by broken molecules of carbon, nitrogen, oxygen and hydrogen such as CH, NH, NH_2, CN and OH, and also C_2 and C_3. What are the parent molecules that split up to produce these unstable radicals? Ammonia (NH_3) and methane (CH_4) are prime suspects, but the suspicion has not yet been confirmed. There is much doubt about the parent molecule for CN. Could it be cyanogen gas (C_2N_2)? Or hydrogen cyanide (HCN)? Or possibly some even more exotic molecule?

Dust Tails and Ion Tails

Regardless of the answers to such questions, it is now understood that the coma of a comet shines with sunlight scattered by dust or with sunlight reradiated by fluorescent gas, usually with both. The tail of a comet is created by another action of the sun. Comet tails, like comet heads, have a gaseous component and a dusty one. For dust tails the action of the sun is uncomplicated: the radiation pressure of sunlight pushes the dust particles out of the coma. Following the laws of motion for orbiting bodies, the dust particles lag behind the coma as they stream away from it; therefore they form a curved tail that can be rich in detail.

Most comets, particularly the brightest, display a huge tail that is only slightly curved. Like the gas in the coma, these tails shine by fluorescence. The molecules responsible for the radiation, however, are ionized, that is, electrons have been removed to leave molecules with a positive electric charge. In such ion tails we find ionized carbon monoxide (CO^+), carbon dioxide (CO_2^+), nitrogen (N_2^+) and the radicals OH^+ and CH^+, but no un-ionized molecules or radicals. Sunlight can ionize some of the molecules, but what pressure can be responsible for pushing them back into space with forces sometimes greater than 1,000 times the gravity of the sun?

The question of how the ion tails are made was long a mystery and has been solved only in the era of space exploration. Space probes have sent back data showing that the sun continuously ejects

COMET KOHOUTEK was photographed on January 11 with the 42-centimeter (16½-inch) Schmidt telescope of the Catalina Observatory on Mount Lemmon in Arizona. The photograph was provided through the courtesy of R. B. Minton of the Lunar and Planetary Laboratory of the University of Arizona, who made numerous exposures of Comet Kohoutek with the instrument. This exposure was made by Steven Kutoroff; it lasted for 10 minutes between 01 : 59 and 02 : 09 Universal Time. The diagonal streak running across the photograph at lower left was made by an artificial satellite; the waxing and waning brightness of the streak is due to the fact that the satellite is irregular in shape and reflects more or less sunlight as it tumbles in its orbit. The photograph was made after perihelion, the comet's closest approach to the sun, so that here the comet is moving in roughly the same direction as that in which the comet's tail is streaming away from its head. Observations of the comet at radio wavelengths show that it contains the "exotic" molecules methyl cyanide (CH_3CN) and hydrogen cyanide (HCN). Since molecules of this kind form in interstellar clouds, their presence supports the hypothesis that comets originate in such an environment.

a million tons of gas per second moving at a radial speed of 250 miles per second. This solar wind, which has a temperature of a million degrees, drags with it chaotic magnetic fields. The fields are carried by currents of electrons in the gas, which is almost completely ionized. Nearly a decade before the first space probe Ludwig F. Biermann of the Max Planck Institute for Physics in Göttingen demonstrated that something like the solar wind was needed to account for the ion tails of comets [see "The Tails of Comets," by Ludwig F. Biermann and Rhea Lüst; page 39]. Although the solar-wind theory of ion tails is not yet very precise, it indicates that two processes couple the solar wind to the cometary gas.

First, the high-energy electrons in the solar wind ionize the molecules in the coma (along with the solar radiation). Second, the solar wind gives rise to a bow wave around the coma. The chaotic magnetic fields now act as a magnetic rake that selectively carries the ions away from the coma, leaving the unionized molecules and atoms unaffected. The force of the solar wind on the ions

can accelerate them to velocities of several tens of miles per second, so that changes in an ion tail can be seen at distances of many millions of miles on a time scale as short as half an hour.

John C. Brandt of the Goddard Space Flight Center of the National Aeronautics and Space Administration has explained the beautiful curvature of these great tails. It results from the transverse motion of the comet at some tens of miles per second across the movement of the solar wind blowing radially from the sun. The ion tails interact with the high-velocity solar wind in the same way that the smoke rising from a smokestack interacts with moving air to produce a graceful billowy arch on the earth.

Cometary Debris

Comets strew debris behind them in interplanetary space. Some of it is seen from the earth as the zodiacal light, which is visible as a glow in the eastern sky before sunrise and in the western sky after sunset. (It is brightest in the Tropics.) Much of the zodiacal light near the plane of the earth's orbit is sunlight

scattered by fine dust left behind by comets. Under ideal observing conditions cometary dust also appears as the Gegenschein, or counterglow: a faint luminous patch in the night sky in a direction opposite that of the sun. Comets need to contribute about 10 tons of dust per second to the inner solar system in order to maintain this level of illumination. Over a period of several thousand years the particles are gradually broken down by collisions with other particles, or are blown away by solar radiation.

In addition to such fine material, comets distribute larger solid particles along their orbit. We see these meteoroids as meteor showers when the earth passes near the orbit of a regular comet. The meteoroids enter the upper atmosphere at speeds as high as 40 miles per second, and atmospheric friction releases the kinetic energy of the object in a short-lived flash of light. The energy released per gram of the meteoroid's weight far exceeds the energy efficiency of the most powerful man-made explosives. Thus an object the size of a pea can create a substantial meteor trail.

No meteoroid associated with a com-

HALLEY'S COMET was photographed in May, 1910, from the southern observing station of the Lick Observatory on Cerro San Cristóbal in Santiago, Chile. Since it has an orbital period of 76 years, it is due for its next return in 1986. Tail of this comet has negligible amount of dust.

COMET AREND-ROLAND was one of two bright comets seen in 1957; the other comet was Comet Mrkos (*see illustration on opposite page*). Photograph was made with the 48-inch Schmidt telescope on Palomar Mountain on April 30.

etary orbit, however, has ever been known to reach the ground, so that it is a very real question whether or not comets contribute any of the meteorites we collect in our museums. In fact, studies of meteors with cometary orbits demonstrate that the cometary debris is extremely fragile. Even though the spectra of such objects indicate that their main constituents are iron, magnesium, silicon and other earthly elements, the material is loosely packed. The meteoroids have an average density of less than half the density of water and can easily be crushed between the fingers. On entering the atmosphere at high speed the material is ripped apart and vaporized. It seldom, if ever, lands in pieces larger than fine dust. Some of the dust, however, can be collected as it floats slowly down through the atmosphere.

Gravel Bank v. Dirty Snowball

How can we be sure that the invisible nucleus of a comet is actually a dirty snowball? Until 1950, when I proposed the icy-nucleus model, the generally accepted conception was that the cometary nucleus was some kind of gravel bank flying through space. It was assumed that the cometary gases had been absorbed in the solid particles and were released by solar energy as the comet approached the sun. Since it was known that comets contribute to interplanetary space solid particles up to at least the size of a grape, it seemed reasonable to assume that such particles were the primary constituent of a comet and that the gases were of secondary importance.

The gravel-bank theory of the cometary nucleus, however, completely fails to pass three critical observational tests. Each of these tests is in itself fatal to the gravel-bank theory but is satisfied by the icy-nucleus hypothesis. The three tests tell us so much about the nature of comets that they merit some elaboration.

The first test involves the fact that comets that come remarkably close to the sun are not destroyed. In 1965 we had an opportunity to observe Comet Ikeya-Seki, which passed the sun at a distance only a third of the sun's diameter away from the sun's surface. At that distance the solar heat is so intense that it would melt and vaporize practically every common substance. There was plenty of time for earthly objects a foot or more in diameter to be boiled away as the comet passed through this solar inferno. A loose gravel bank would have been completely vaporized as the tidal forces due to the sun's gravity were tearing it apart. The intense solar wind would then have removed the resulting gas and completed the destruction of the comet. Comet Ikeya-Seki withstood the holocaust practically unscathed. Indeed, it is one of a group of eight such comets. They apparently had a common ancestor that broke up into separate comets, presumably because of the tidal forces due to the sun's gravity.

Sun-grazing comets pass so close to the sun as to be within the well-known limiting distance established by E. Roche of France in the middle of the 19th century. Roche showed that when a small fluid body passes a large body within 2½ times the radius of the large body, the gravitational force on the near side of the small body is much greater than it is on the far side. The force is enough to pull the body apart against its own simple gravitational cohesion. Only a body

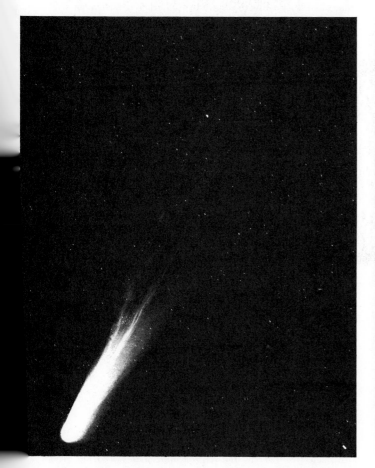

COMET MRKOS, visible in August, 1957, clearly shows that a comet has two kinds of tail and also that the form of each tail changes with the passage of time. The straight tail to the upper left in both photographs that has the appearance of cigarette smoke caught in a breeze is composed of ionized gases. Fainter, smooth tail curving off below it is composed of dust. The photograph at left was taken on August 22, the one at right five days later. Both were made with the 48-inch Schmidt telescope on Palomar Mountain.

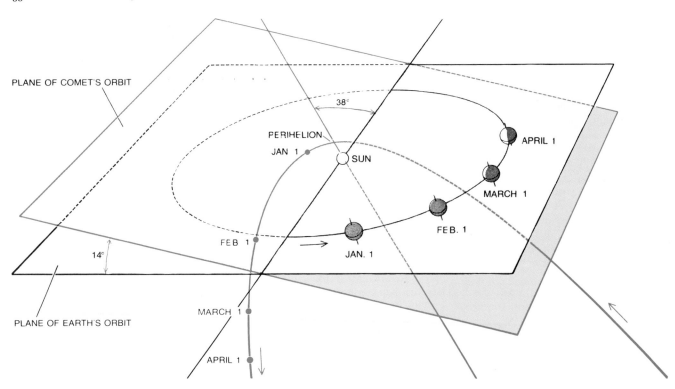

PLANE OF COMET'S ORBIT

38°

PERIHELION

JAN 1

SUN

APRIL 1

MARCH 1

FEB. 1

FEB 1

JAN. 1

14°

PLANE OF EARTH'S ORBIT

MARCH 1

APRIL 1

ORBIT OF COMET KOHOUTEK (*dark color*) is shown with respect to the orbit of the earth (*black*). Kohoutek's orbit is an extremely elongated ellipse whereas the earth's orbit is nearly a perfect circle. The plane of the comet's orbit (*light color*) is inclined at an angle of only 14 degrees to the plane of the earth's orbit (*gray*). At perihelion Comet Kohoutek was some 13 million miles from the sun. Dates along the two orbits indicate the position of the comet and the earth with respect to each other at various times.

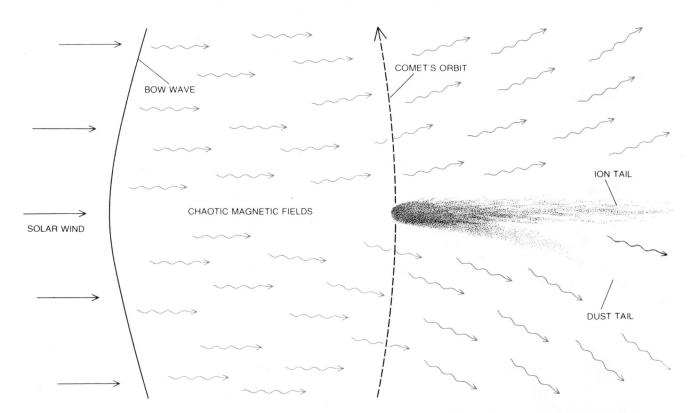

BOW WAVE

COMET'S ORBIT

ION TAIL

CHAOTIC MAGNETIC FIELDS

SOLAR WIND

DUST TAIL

ION AND DUST TAILS of a comet are created by two different processes. The ion tail is formed by a double process. First high-energy electrons in the solar wind (gases ejected from the sun at high speed) ionize the molecules in the coma of the comet, stripping them of electrons and leaving them positively charged. Second, the solar wind gives rise to a bow wave around the coma. Chaotic magnetic fields within the solar wind act as a magnetic rake that selectively carries the ionized molecules away from the coma at high speeds. Dust tail is formed by pressure of sunlight pushing the dust particles out of the coma. Tail is curved because the dust particles follow the laws of motion of orbiting bodies and lag behind the coma as they stream away at relatively low speeds.

with more than gravitational cohesion can withstand the tidal effects within Roche's limit. Even a body with such cohesion might split up along cracks or in other areas of weakness. We can therefore conclude that the nuclei of comets, at least comets of the sun-grazing family, have some internal strength but not always enough to protect them from the tidal disruption of a particularly close encounter with the sun. These facts alone eliminate the gravel-bank model of the cometary nucleus, unless one wishes to postulate that various comets differ completely in their basic structure.

The Second Observational Test

The second observational test fatal to the gravel-bank model is the persistence of some of the periodic comets. About 100 comets of the 600 so far observed move in orbits with periods of less than 200 years. All of those with periods of less than 30 years move around the sun in the same direction as the planets, and the planes of their orbits are inclined an average of 12 degrees to the plane of the earth's orbit. Some of these comets have completed many revolutions around the sun during recorded history. The painstaking records of the Chinese show that Halley's comet has appeared near the sun 29 times at intervals of between 76 and 77 years. The first well-observed passage was in 239 B.C. and the most recent was in 1910. In 1986 we shall see Halley's comet again.

From the observations made in space by the Orbiting Astronomical Observatory satellites (OAO) we know that bright comets lose tons of gas per second over periods of months as they come close to the sun. On the gravel-bank theory one would expect a single passage to remove a major fraction of the gas that the particles in the gravel bank have absorbed. The comet should then drastically diminish in brightness on subsequent returns. Halley's comet has probably lost a cubic kilometer of ice during its known history, perhaps even more. Interplanetary space contains so little matter that the gravel bank could not absorb enough gas to replenish its supply during the part of its lifetime spent at great distances from the sun. Thus a comet must have a huge reservoir of ice and solids to fuel its spectacular performance each time it passes close to the sun. For Halley's comet an icy nucleus some five miles or more in diameter is clearly indicated. The comet need lose only a fraction of 1 percent of its total substance on each return.

Now we come to the final test. A mys-

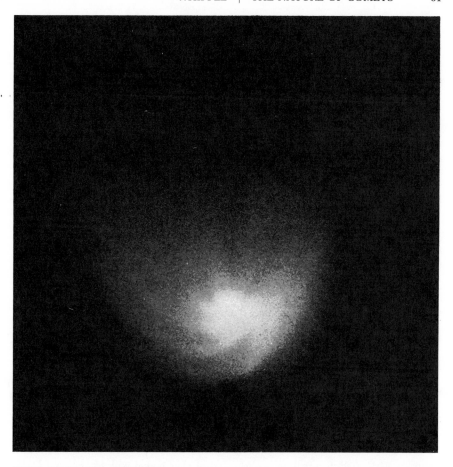

COMA OF COMET BENNETT, a bright comet visible during the spring of 1970, shows a detailed spiral pattern in this photograph made by Minton and Stephen M. Larson on March 28, 1970. They made 10 photographs ranging in exposure time from four to eight seconds, using the 61-inch reflecting telescope of the Catalina Observatory. This picture is a composite of the 10 images. Minton and Larson interpret the spiral pattern as evidence that the icy nucleus is rotating on its axis with a period of about a day and a half as it orbits the sun.

terious idiosyncrasy of comets is that they defy Newton's law of gravity. In 1819 J. F. Encke of Germany studied the motion of a short-period comet discovered in 1786. To his surprise he found that the comet, which has the shortest cometary period known (3.3 years), persisted in returning at each revolution about 2½ hours too soon. Only 2½ hours in 3.3 years may seem like a trivial deviation, but the accuracy with which astronomical positions can be measured places it well outside the possible errors of measurement and calculation.

Nongravitational Motion

Comet Encke still persists in returning sooner than predicted on each revolution. Brian G. Marsden and Zdenek Sekanina of the Smithsonian Astrophysical Observatory find, however, that the effect has slowly diminished to about 10 percent of the deviation it exhibited 150 years ago. Even Halley's comet shows maverick tendencies in its motion. T. Kiang of the Dunsink Observatory in Dublin finds that during its past 11 apparitions the comet persists in arriving late by an average of 4.1 days. Marsden and Sekanina show that of 20 short-period comets all but two show deviations from Newtonian motion, half being accelerated and half retarded.

The gravel-bank model completely fails to account for this nongravitational motion among comets. Until space probes had proved that interplanetary space was a nearly perfect vacuum, one might have assumed that there was a drag caused by some resisting medium. We now know that there is not enough matter in space to produce the required retarding effect. Moreover, even if there were such a drag, there would be no way for it to push certain comets forward in their orbits.

The icy nucleus provides a simple explanation for the wayward motions of comets. In the glare of the solar radiation the ices evaporate and send out molecules at speeds of several hundred feet per second. The molecules are ejected on the sunny side of the nucleus and

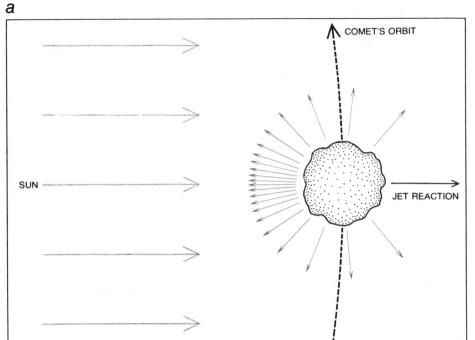

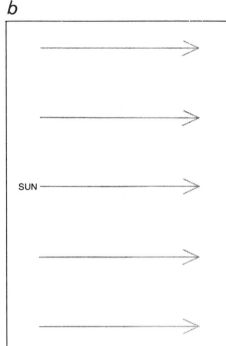

COMETS DEFY LAW OF GRAVITY by speeding up and slowing down in their orbits. The radiation from the sun causes the ices to sublime (evaporate) and send out molecules at speeds of several hundred feet per second. The molecules (*color*) are ejected on the sunny side of the nucleus and generate a jet reaction. This force pushes the comet away from the sun (*a*). If the nucleus is rotating, the jet will be displaced along the direction of the comet's orbital motion. If the nucleus is rotating in the same direction as its motion

generate a jet reaction, a real force pushing the comet away from the sun. If such a force is to change the period of a comet, however, it must be directed along the path of a comet, either fore or aft. Now, suppose the comet nucleus is rotating. (And we have yet to observe a celestial object that does not rotate to some extent.) The jet will be displaced by the rotation, that is, there will be a delay in the ejection of gas from any point on the surface as it rotates past the point directly facing the sun.

The delay is similar to the lag of the seasons on the earth: the summer temperature in the Northern Hemisphere is highest late in July, not at the time of maximum solar heating on June 21. If a comet is rotating in the same direction as its motion around the sun, the delayed jet action will have a forward compo-

nent. The comet will drift outward from the sun in its orbit, and its orbital period will be increased. As a result it will show up later than predicted.

On the other hand, if a comet rotates in a direction opposite the direction of its motion around the sun, the jet action will have a backward component. The comet will "feel" a drag force, and it will drift slowly toward the sun in its orbit. Its period will be reduced, and it will show up earlier than predicted. Calculations I made in 1950 demonstrate that for a comet nucleus that has a diameter of the order of a very few miles, and that utilizes solar energy fairly efficiently in subliming its ices, the calculated force is adequate to produce the changes observed in the arrival times of the periodic comets. If the comets' axes of rotation are randomly distributed, we

should expect about half of them to be accelerated and half retarded. That, of course, is the case.

The Water Component

I later found that, if the icy-nucleus model is to account for the changes in the periods of comets, it would have to eject gas at a rate that is considerably higher than what can be deduced from the intensity of the lines in the spectrum of a comet at the wavelengths of visible light. Otherwise the jet force would not be strong enough to exert an adequate force on the larger nuclei. Then Biermann showed that if ordinary frozen water is the major component of icy nuclei, bright comets should be surrounded by huge clouds of hydrogen and the hydroxyl radical arising from dissociated

COMETS COME FROM A HUGE CLOUD of isolated icy nuclei, according to the theory put forward by J. H. Oort. The Oort cloud is gravitationally a part of the solar system but extends out some 50,000 astronomical units in radius. (One astronomical unit is the distance from the earth to the sun.) The open circle at the right represents the diameter of the solar system. The small dots repre-

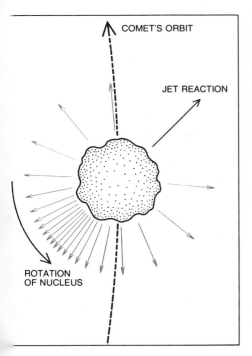

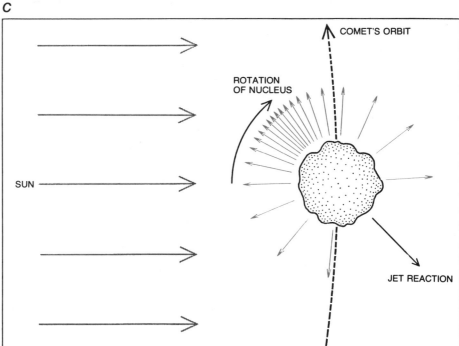

around the sun (*b*), the delayed jet reaction will push the comet forward in its orbit. The nucleus will drift outward from the sun, its orbital period will increase and it will show up later than predicted. If the nucleus rotates in a direction opposite to its orbital motion around the sun (*c*), jet reaction will have a component of motion opposite to comet's direction of travel. Comet will experience a drag force and will drift inward toward the sun. Its orbital period will be reduced and it will show up earlier than predicted.

water molecules. A hydrogen atom isolated in deep space sends out almost all its fluorescent radiation not in the visible region of the spectrum but in the very far ultraviolet at the spectral line designated Lyman alpha.

In 1970 Arthur D. Code and Charles F. Lillie turned the first Orbiting Astronomical Observatory to observe Comet Bennett. They discovered a large cloud of hydrogen radiating at the wavelength of the Lyman-alpha line. They also found that the hydroxyl radiation in the far violet, which is barely detectable on the earth, is much stronger than had been suspected. The gas that was radiating at these wavelengths was what was missing from my calculations. Moreover, there seem to be approximately equal numbers of hydrogen atoms and hydroxyl radicals, confirming the suspicion

that frozen water is a major constituent of a comet nucleus. Thus the observations verified Biermann's hypothesis and the basic conclusions drawn from the icy-nucleus model. The observations were confirmed by J. Blamont and J. Bertaux of the National Center of Space Studies in Paris; since then large quantities of hydrogen have been detected in three other comets, including Comet Kohoutek.

The Origin of Comets

Now that we know the nature of a comet nucleus, what might its detailed chemical and physical structure be? Could that structure disclose the origin, evolution, lifetime and final disposal of comets? It is easy to make comets, at least sitting at one's desk. All that is

needed is to collapse and cool a cloud of interstellar gas and dust. There are many collapsed clouds in the central plane of our galaxy, where new stars and probably new solar systems are now evolving. One of these stellar incubators is the region of the Great Nebula of Orion.

When the temperature in a highly collapsed cloud falls low enough, say to a few tens of degrees above absolute zero, all earthly substances such as water, ammonia and methane will freeze into solid particles. The molecules observed in comets are simple combinations of carbon, nitrogen, oxygen and hydrogen. If comets originate in a typical interstellar cloud of our galaxy that has abundances of atoms much like the abundances in the sun, these four elements are by far the most prevalent ones capable of form-

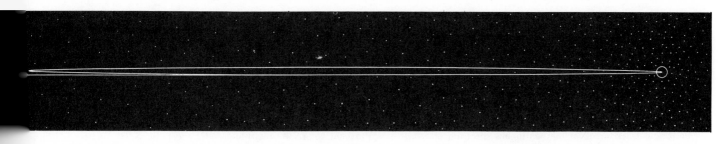

sent the comet nuclei in this section of the Oort cloud. A star passing through the outer regions of the cloud (approximately 10 feet to the left off the page on the scale of the drawing) will disturb the motions of some of the nuclei just enough for them to enter the solar system. Ellipse shown is the correct scale for orbit calculated for Comet Kohoutek: 44 astronomical units wide and 3,600 long.

ing compounds. Although a quarter of the sun's mass is helium, helium and the other noble gases will not combine chemically with other elements, nor will they freeze out at the expected temperatures and densities.

Are the comets we observe actually created by this process? The solar system's comets form a huge cloud, gravitationally a part of the solar system but extending out several thousand times farther than the outermost planet (Pluto). As J. H. Oort of the University of Leiden showed in 1950, a star passing near the outer reaches of this cometary cloud will disturb the motions of some comets just enough for them to enter the inner part of the solar system where we can observe them. Otherwise they remain deep-frozen, perhaps since the sun and planets were formed 4.6 billion years ago. Comet Kohoutek appears to be one of the "new" comets entering the planetary region for the first time; thus it may be a sample of the primordial material of the solar system.

Interstellar Cloud v. Solar Cloud

Where did the comets in the great Oort cloud originate? A. G. W. Cameron of the Harvard College Observatory has recently suggested that they formed from fragmentary interstellar clouds at the time when a larger cloud spun down into a disk and formed the solar system. The comets would then be created moving in huge orbits like those of new comets such as Comet Kohoutek before they are disturbed by a passing star. It is not yet certain whether or not the ices and solids needed to form sizable comets could aggregate in a fragmentary interstellar cloud. On the other hand, we are certain that comets must have been formed in huge quantities at the outer edges of the solar system.

The planets Uranus and Neptune have a mean density about twice the density of water, just what one would expect if they were aggregates of literally hundreds of billions of comets. The terrestrial planets (the earth, Mars, Venus and Mercury) must have been formed from large numbers of planetesimals, or small earthlike aggregates. The difference between Uranus and Neptune and the terrestrial planets appears to depend simply on the temperature in the two regions of space at the time they were formed. One would expect that in a great collapsed interstellar cloud rotating as a disk the temperature would remain higher near the burgeoning sun than in the outskirts of the disk. Apparently water could not freeze inside the orbit of Jupiter but could freeze outside the orbit of Saturn. (Jupiter and Saturn themselves, like the sun, are made mostly of hydrogen and helium, compressed by the great mass of the planets to roughly the density of water.) Therefore we can be fairly sure that comets formed beyond the orbit of Saturn. Icy grains and grains of earthlike material formed in the cooling disk of the solar cloud, accumulated into comets and finally aggregated into Uranus and Neptune. After Uranus and Neptune became massive the remaining comets were disturbed by the gravitational attraction both of the new planets and of Jupiter and Saturn, and they were swung into the huge orbits of the Oort cloud.

The process would have been extremely wasteful. Many comets would have been captured by Jupiter and Saturn. Many would have decayed into smaller orbits in the inner solar system as they do today. Many would have been lost into the depths of the galaxy. No one has yet shown, however, that comets formed within the orbits of the present outer planets could actually move outward into the great Oort cloud, nor, as we have seen, has it been proved that fragmentary interstellar clouds could condense and aggregate into comets.

Resolving the Dilemma

To determine whether comets formed in the cloud that gave rise to the solar system, in fragmentary interstellar clouds or in some other kind of environment, clues are needed both from new theories of comets and from new observations. Suppose comets did form in fragmentary interstellar clouds. Then each interstellar dust grain would be a nucleus for the condensation of the various ices. The ices might consist largely of the "exotic" molecules now being discovered by radio astronomers in clouds of interstellar gas and dust. These molecules include formaldimine (CH_2NH), methyl alcohol (CH_3OH), methyl cyanide (CH_3CN), hydrogen cyanide (HCN) and more than 20 others, mostly composed of elements that are abundant

SPECTRUM OF A COMET HAVING MUCH GAS in relation to dust shows many distinct spectral lines composing one molecular band. The spectrum, which is of Comet Ikeya (visible in 1963), spans the blue wavelengths from 3,900 angstroms (*left edge*) to 4,100 angstroms (*right edge*). Since this spectrum is a negative print, the dark lines are actually bright emission lines. Almost all the features are from the carbon molecule C_3. The featureless horizontal streak in the center is weak spectrum of reflected sunlight.

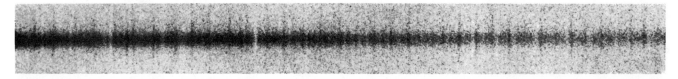

SPECTRUM OF A COMET HAVING MUCH DUST in relation to ionized gas shows few and weak distinct spectral bands and lines. The horizontal black streak of the continuous spectrum of reflected sunlight, however, is very strong. The spectrum, which is of Comet Mrkos, is also crossed by absorption lines that show up light on this negative print. These indicate the wavelengths at which the dust absorbs light from the solar spectrum. Spectrum runs from approximately 4,880 angstroms (*left edge*) to 5,100 angstroms (*right edge*). Both this spectrum and the one at the top are from Jesse L. Greenstein of the California Institute of Technology.

in comets: carbon, nitrogen, oxygen and hydrogen.

Are some of these exotic molecules the parent molecules of the simple radicals we see in comets? If they are, then comets probably formed in fragmentary interstellar clouds. If, on the other hand, comets originated within the orbits of the outer planets, one would expect the parent molecules to be the more stable ones such as ammonia and methane. As Pol Swings and Armand Delsemme of the Institut d'Astrophysique in Belgium have suggested, even though the temperature in the solar cloud may never have been low enough for methane to freeze out (some 20 degrees above absolute zero), such molecules were probably trapped in water snow.

The two possible birthplaces of comets might also give rise to marked differences in the structure and behavior of comets. Solar-cloud comets made of dust grains surrounded by ices could be expected to disintegrate to nothing as their substance sublimes. On the other hand, comets formed in the outer solar cloud might well have developed an earthlike core if the temperature of the cloud's great disk first rose during its collapse and then slowly fell. An old comet of this type might lose its outer shell of ices, slowly become inactive and finally appear starlike instead of fuzzy in our telescopes, looking very much like an asteroid. Indeed, many astronomers believe the small asteroids observed to cross the earth's orbit are really old comet nuclei.

A Space Probe to a Comet

The best way to answer these questions about comets, and thereby questions basic to an understanding of the origin of the solar system, is to send space probes to the nucleus of a comet and to the surface of an asteroid. Such space probes, which would actually be unmanned observatories, are completely feasible today, both technologically and scientifically. They would be less expensive than several of the planetary probes that have already proved to be so successful.

NASA is currently studying the possibilities of such space missions to comets and asteroids. The earliest mission now being considered would be to Comet Encke in 1980. Meanwhile the interest in comets renewed by Comet Kohoutek should result in new information of prime importance for our understanding the true nature of the most primitive bodies in the solar system.

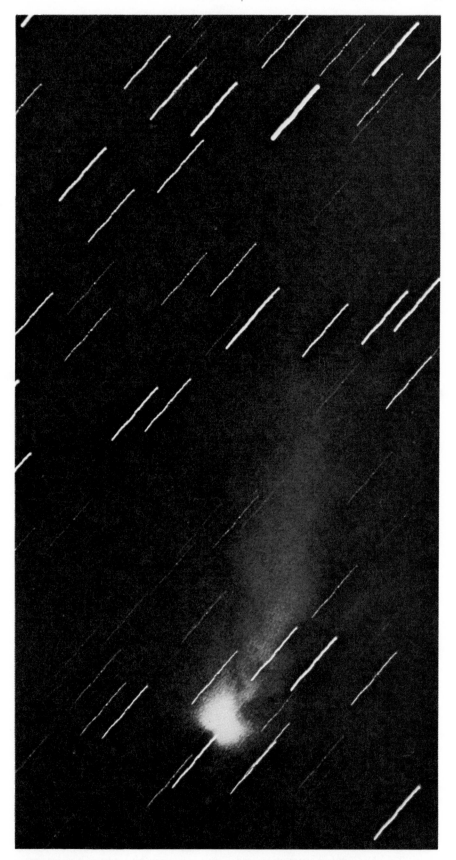

COMET HUMASON, a relatively faint comet visible in 1960, was unique in that its spectra showed almost no dust at its great perihelion distance of some four astronomical units from the sun. The blue color is due to emission from ionized molecules such as carbon monoxide (CO^+), molecular nitrogen (N_2^+), the hydroxyl radical (OH^+), cyanogen (CN) and CH^+. Since the telescope (the 48-inch Schmidt telescope on Palomar Mountain) was following the comet's motion, the images of the background stars appear as streaks.

7

The Spin of Comets

by Fred L. Whipple
March 1980

*Puzzling changes in the orbital period of Comet Encke,
which circles the sun every 3.3 years, can be attributed
to the rotation of its icy nucleus and the thrust of
gases evaporated from it*

A comet, like an atom, has a nucleus. The nuclei of comets, however, have been more elusive than those of atoms. The atomic nucleus was first identified in 1911, some 40 years before I was able to show that comets have a nucleus (which I likened to a dirty snowball). There was a somewhat greater interval between the introduction of the concept that atomic nuclei spin and the first measurements of a few rotation periods for comet nuclei in 1977. The spin axes of atomic nuclei have long been known to gyrate when they are subjected to a magnetic field. Here I shall describe the gyration of the spin axis of a comet nucleus.

In the study of the nuclei of atoms and the nuclei of comets two similarities stand out: invisibility and relative size. No one can see an atomic nucleus, and very rarely, if ever, can anyone see the true solid nucleus of a comet, even with the aid of a large telescope. The nucleus of an atom is about a hundred-thousandth of the diameter of the atom. A similar ratio holds for the nucleus of a comet with respect to the comet's gassy and dusty coma, or head, which is what is seen or photographed.

The physicist has had an advantage over the astronomer in the study of a body so much smaller than the object of which it is a part. Atoms are vastly more abundant than comets. Moreover, their nuclei have an electric charge and a magnetic polarity (because of their spin). Where the physicist can probe an atomic nucleus whenever laboratory apparatus is available, the astronomer must wait patiently for a distant view of one of the few comets that swing by. The astronomer's handicap would be greatly eased, however, if a proposed space mission were viewed favorably by Congress. The National Aeronautics and Space Administration has developed detailed plans for a spacecraft rendezvous with Halley's comet in November, 1985, followed by a prolonged visit three years later to a much less famous comet: Tempel 2. In a period of budgetary austerity funding for the mission is current-ly uncertain. Meanwhile astronomers do the best they can with the tools they have at hand.

The most valuable comet for the study of the comet nucleus is the one with the shortest orbital period (3.3 years), discovered in 1786 by Pierre Méchain but named for the German mathematician and physicist Johann Franz Encke. In 1819 Encke studied the motion of the comet and concluded that its motion deviated "wildly" from predictions based on Newton's law of gravitation. He found that its period was getting about $2\frac{1}{2}$ hours shorter with every revolution around the sun.

At that time the decrease was ascribed to a resistant medium in space. It was even suggested that the medium might be the "luminiferous aether" supposedly needed to carry light waves through otherwise empty space. As late as 1950 an elementary textbook of astronomy stated that "a comet appears to be a swarm of relatively small and widely separated solid bodies held together loosely by mutual attraction." It is only such a "flying gravel bank" that would meet resistance in space. Larger solid bodies such as asteroids and planets show no such effect, and neither would a solid comet nucleus.

By 1868, however, the rate at which Comet Encke's period was getting shorter had mysteriously decreased and thereafter continued to do so. Today the decrease in period amounts to only a few minutes per revolution. By 1950 several other comets had also been observed to follow a pattern of erratic deviation from pure gravitational motion, some with decreasing periods like the period of Comet Encke but some with increasing ones. Halley's comet, for example, returns about four days late after each of its 76-year revolutions around the sun.

From such observations I developed my theory of the comet nucleus as a ball of dirty ice. Let us imagine that the ice ball is also rotating on its axis so that it has a day and a night. Points on its surface alternately face toward the sun and away from it. Surface ices will evaporate (more correctly, sublime) much faster when they are heated by the sun than they will at night, when they are exposed to the near absolute zero of black outer space. In the "morning" of each comet "day" the frigid ices are gradually warmed by the sun and reach a maximum temperature toward the late afternoon. The outflowing gases arising from the sublimating ices generate a reactive jet force that pushes the nucleus of the comet at an angle equal and opposite to the angle the thrust of maximum sublimation makes with a line drawn to the sun.

If the nucleus is rotating in a direction opposite to that in which it revolves around the sun, the afternoon jet thrust will oppose the comet's motion, thereby slightly reducing its orbit and hastening its arrival at perihelion: its closest approach to the sun. The reverse direction of rotation will accelerate the comet, thereby expanding its orbit and delaying its arrival at perihelion.

A reverse sense of rotation must therefore hold for Comet Encke and for about half of some three dozen other comets whose orbital-period changes have been determined by Brian G. Marsden and Zdenek Sekanina of the Smithsonian Astrophysical Observatory and Donald K. Yeomans of the Jet Propulsion Laboratory of the California Institute of Technology. A comet need lose only a small fraction of 1 percent of its total mass per revolution to account for the observed changes in period; most comets can therefore survive a great many revolutions around the sun before they exhaust their deep-freeze supply of ices.

Why, however, should the period change for Comet Encke be so drastically reduced in less than 200 years from $2\frac{1}{2}$ hours per revolution to a few minutes? The popular explanation has been that Comet Encke is slowly losing mass and that the jet force has simply decreased over the period of observation. This explanation is somewhat too pat, even though Comet Encke may not be quite as bright as it was 200 years ago. (The brightness observations are open to question.) Still, the surface of an aging comet might accumulate dusty material that would cover up the volatile ices and so reduce the jet effect. On the other hand, why did the nongravitational change in period rise to a maximum in about 1810 and then fall?

Sekanina, my colleague at the Smithsonian Astrophysical Observatory, and I had long suspected that the real cause might lie in the changing spin axis of the nucleus. This could affect the geometry of the jet forces and so account for the phenomenon. But how could one find the spin axis of a tiny comet nucleus buried inside the coma, brilliant with fluorescing gases and dust-scattered sunlight? At about the same time we both came to approximately the same

method, although we approached it from entirely different points of view. For the sake of brevity I shall describe only Sekanina's method, which applies directly to Comet Encke.

The delay time in the sublimation of ices on a rotating comet nucleus will cause the escaping gas and dust to form an asymmetrical or fan-shaped jet on the afternoon side of the nucleus. Generally the jet will point in a direction

COMET ENCKE, the comet with the shortest period of revolution around the sun (3.3 years), was photographed on November 22, 1937, 35 days before perihelion (the point at which the comet comes closest to the sun). Comet Encke was then 131 million kilometers from the sun and 43 million kilometers from the earth. Its distance from the sun at perihelion is about 51 million kilometers. Close inspection of such photographs discloses that the invisible solid nucleus of the comet is emitting a jet of volatilized gases at an angle that is not on a line with the sun. Normally the gas and dust surrounding a comet nucleus is formed into a tail pointing directly away from the sun. The asymmetrical jet is evidence that the nucleus of Comet Encke is spinning. Although Comet Encke is never a spectacular object, its frequent apparitions make it probably the most thoroughly studied comet. The photograph was made by George Van Biesbroeck with a 24-inch reflector at the Yerkes Observatory. Bright streaks in the background are stars in apparent motion as telescope tracked the moving comet.

different from that of the tail of the comet, which points directly away from the sun. The coma will be elongated or will be displaced from the bright region centered on the nucleus. Observers frequently note such asymmetries in the head of comets and record the direction of the jets and fans.

To determine the spin axis of a comet nucleus Sekanina wrote an elaborate computer program that provides for each comet observation a printout of a three-dimensional table. The table gives the position on the celestial sphere toward which a jet points for all possible directions of the spin axis and all possible angles of the lag in sublimation. From the table Sekanina can pick out the possible ranges of spin-axis direction and lag angle to fit each observation. The range of solutions narrows as he compares more observations.

Applying this method to four short-period comets, Sekanina discovered that as of 1947 the spin axis of Comet Encke was tilted five degrees with respect to the plane of the comet's orbit around the sun. At perihelion the spin axis made an angle of 25 degrees with respect to a line

between the sun and the comet. The lag angle of the jet action was 45 degrees. Clearly the change of period caused by the jet action is complicated to calculate because the jet force varies tremendously both in amount and in direction as the comet travels around its orbit. Only by a numerical integration can one calculate how the orbital period should change. Without the modern electronic computer the task of integration would take a lifetime.

Sekanina and I next assumed that the nucleus of Comet Encke is not a sphere but an oblate spheroid, something like a doorknob. A nearly rigid rotating body in space will quickly adjust its spin axis and mass so that the spheroid rotates stably about its shortest axis. This is recognized in physics as the axis with the maximum moment of inertia. The geometry immediately indicates that the jet force perpendicular to the surface will rarely pass through the center of the nucleus. The result is an overturning force that tends to tip the pole of the nucleus in one direction or the other. It is well known, however, that a spinning top or a gyroscope does not respond to

the pull of gravity by falling over. The pole of the top turns in a direction 90 degrees with respect to a plane defined by the force and the axis. Hence the oblate comet nucleus precesses like the earth, which is tipped by the pull of the moon and the sun on its equatorial bulge. For the earth the precession is extremely slow, having a period of about 25,000 years.

With the geometry in order for our computer programs, Sekanina and I devised a theoretical model of the jet force, which for comets increases faster near the sun than the inverse square of the distance to the sun. We were guided by the many observations showing that Comet Encke has a very peculiar light curve. Instead of brightening rather uniformly as it comes closest to the sun at perihelion and then dimming more slowly after perihelion, as most comets do, Comet Encke is brightest very nearly at perihelion and then almost immediately gets much dimmer than it was at the same distances before perihelion.

By 50 days after perihelion it is three magnitudes, or 16 times, dimmer than it was 50 days before it. The observations compel one to conclude that one polar hemisphere of the comet is much more active than the other. This conclusion gives an additional numerical relation for the jet force, determined by the latitude on the nucleus at which the sun is overhead at each moment in the comet's orbit. When the sun shines directly on the pole that approximately faces the sun just after perihelion, the comet is nearly three times fainter than it is when the sun shines on the opposite pole, a surprising fact demanded by the observations.

Separately we programmed two different computers, working with two different coordinate systems to calculate both the precession rate of the spin axis and the nongravitational orbital forces. The jet force varied in direction and amount with the actual position of the comet in its orbit from 1786 to 1977, representing 59 perihelion passages. We were gratified with the immediate results. They showed that the pole would sometimes precess as much as one degree a year, which accounts for the strange nongravitational motion of the comet.

After a number of iterations of the computations to fit the nongravitational motion determined by Marsden and Sekanina, we derived only small corrections of about four degrees in the direction of the polar axis and less than one degree in the lag angle of 45 degrees determined earlier by Sekanina. The solution reproduces the curve of observations within the accuracy of the orbital calculations [see bottom illustration on page 61]. Slight discrepancies between the observations and the calculated

CHANGES IN ORBITAL PERIOD of Comet Encke were first recognized in 1819 by Johann Franz Encke, for whom the comet was named. Encke found that the comet returned to perihelion some $2\frac{1}{2}$ hours early after each trip around its orbit. Since about 1830, however, the amount by which the orbital period is getting shorter has declined steadily; it is now only a matter of minutes. Colored curve is a highly smoothed representation of actual observations. White dots mark perihelion passages. The next perihelion passage will be on December 6 of this year.

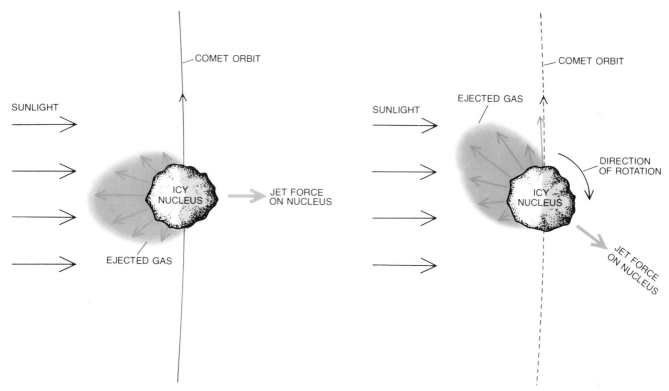

THRUST OF VOLATILIZED GASES can alter the orbit and hence the period of a rotating comet. If the comet nucleus is nonrotating (*left*), surface ices heated by the sun can generate a small but steady jet force that will push the comet radially outward from the sun. Because the outward thrust constantly subtracts a small amount from the acceleration due to the sun's gravity the orbital period will not change from one revolution to the next. In principle it should be possible to tell if the radial jet force exists: a comet subjected to it should occupy an orbit slightly larger than the orbit of a comet in which the force is absent. In actuality the difference in the size of the orbit would probably amount to less than one part in 10⁵ and so would be below the limit of detectability. If the comet nucleus happens to be rotating in a direction opposite to the direction of its motion around the sun (*right*), the thrust of volatilized gases will reach a peak in the "afternoon" of the comet's "day." The resulting jet force will subtract slightly from the kinetic energy of the comet, steadily contracting its orbit. As a result the comet's period will be observed to shorten slightly with each perihelion passage. A comet rotating in a direction opposite to the one at the right will experience an expansion of its orbit and therefore a lengthening of its period. Evidently shortening of Comet Encke's period has been basically due to the comet's rotation in a direction opposite to the direction in which it moves around the sun.

points are caused by the gravitational action of Jupiter. The pull of the giant planet changes the orbital plane of the comet by a fraction of a degree and consequently changes the geometry of the jet action.

When we plotted the pointing direction of the spin axis of the nucleus of Comet Encke on the celestial sphere, we were astonished: in 191 years the pole of rotation of the nucleus has gyrated through more than 100 degrees across the sky. This large gyration completely accounts for the changes in the orbital period. The century-old mystery was solved.

One independent check on our theory remained. Would the calculated spin axis agree with the observed directions of the fans and asymmetrical comas, data that we had not used in developing our solution? To our delight the fit with more than 30 such observations from 1805 to 1904 was completely satisfactory, and the fit with the later observations used originally by Sekanina was greatly improved. The spin axis of Comet Encke has assuredly moved as is indicated by the calculations.

Where will the spin axis of Comet Encke be pointing in future years? Our calculations show that by 1990 the decrease in the period of the comet's revolution will stop and the period will begin to increase again. If the shape of the nucleus is not changed too much by the loss of ices, the spin axis will be almost perpendicular to the plane of the comet's orbit by the year 2200. Then the comet will be spinning like the earth: in the same sense as it revolves around the sun. Such a rotation axis is quite satisfactory for a planet but is eventually unstable for a comet. Since the sun shines mostly on the equatorial regions the nucleus will become spindle-shaped, elongated along its polar axis because of the loss of mass. The body of the comet will then twist around so that further prediction of the gyration is impossible today.

The past history of the spin axis is more interesting and determinable. According to our calculations, the axis was stuck in almost the same position for perhaps hundreds or even thousands of years before A.D. 1700. One pole was pointed nearly along the long axis of the orbit so that almost all the mass loss occurred on that polar hemisphere when the comet was close to the sun. Indeed, that hemisphere is the one we see today as being the most active. Is Com-

et Encke rockier or dustier on one side and icier on the other? Apparently it is. Is this property, however, basic to the structure of the comet? Sekanina and I prefer another explanation, one that does not call for such a drastic nonuniformity in the initial composition of the object.

Comets are known to blow quite sizable pieces of rocky material off into free orbit around the sun. When such comet debris happens to enter the atmosphere of the earth, it is seen as a meteor or a meteor shower. For a given comet at a given distance from the sun there is a limit on the size of the pieces the gas escaping from the ices can blow off. A comet nucleus is only one kilometer to a few kilometers in diameter, but it still has some small gravitational pull. The gas pressure, however, is also small. Sekanina and I suggest some of the particles near the size limit that are ejected from the most active areas of the comet go into long orbital trajectories and finally land on the night hemisphere of the nucleus. If one polar hemisphere never faces the sun except at great distances where the comet is inactive, the dusty or rocky debris slowly blankets that hemisphere during hundreds of rev-

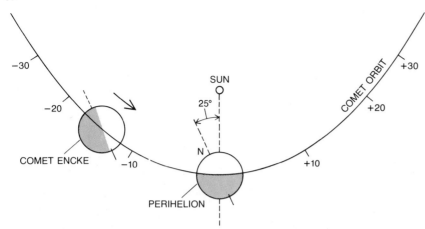

ORIENTATION OF THE SPIN AXIS of Comet Encke in 1947 was determined by Zdenek Sekanina of the Smithsonian Astrophysical Observatory. On the basis of observed asymmetries in Comet Encke's appearance Sekanina concluded that the spin axis was rotated about 25 degrees eastward from a line drawn between the comet and the sun at perihelion. Pole nearest the sun (designated north) was tilted five degrees below the plane of the orbit. Ten-day intervals are marked on the orbit. At perihelion in 1947 comet was 51,017,000 kilometers from the sun.

olutions around the sun. By that time, the gyration having turned the inactive hemisphere toward the sun near perihelion, the blanket of debris insulates the underlying ices of the comet from the solar heat and greatly reduces the comet's activity. This sequence of events could account for the peculiar light curve of Comet Encke.

A space mission to Comet Encke could check this suggestion about the character of the comet nucleus. Otherwise astronomers hundreds of years from now might note that the rocky blanket had finally blown off the dark hemisphere, or perhaps determine that

the nucleus is indeed lopsided with respect to icy and rocky material. The answer to the question is important in reconstructing the evolution of comets and of the entire solar system. Do larger comets have a rocky core? Do some of them finally lose their icy mantle and contribute small asteroids to the inner solar system?

All the evidence of modern astronomy supports the idea that stars like the sun and their planetary systems (if any) originate in collapsing clouds of interstellar gas and dust. The material is extremely cold, only a few degrees above

absolute zero, until it is heated by the collapse to the region where the star is forming. If comets have no rocky core, they must have formed at the edge of the primordial nebula where the gas and dust was never heated. If comets do have a rocky core, the temperature must first have risen enough to sublimate the ices so that the dust in the comet could aggregate into a rocky core. Then the temperature would have had to fall in order to enable the core to collect its icy mantle, still containing considerable dust.

Our calculations provide another new fact about the nucleus of Comet Encke: the ratio of the precession rate of the spin axis to the total nongravitational motion in the orbit. Clearly the precession rate should increase with the oblateness of the nucleus. Moreover, the faster the nucleus spins, the slower the precession rate should be. It turns out that the ratio decreases with the diameter of the nucleus but does not involve the density. Therefore our new ratio is closely proportional to the oblateness of the nucleus multiplied by the period of rotation and divided by the diameter of the nucleus. If we can determine any two of three quantities (oblateness, rotation period and diameter), the third is also determined. There is no hope, however, of directly observing the oblateness without a space mission to the comet, so that one must try to measure or estimate the other two quantities. First consider the diameter.

Comet Encke has never come close enough to the earth for astronomers to see its nucleus, and so there is no direct measure of the diameter of the nucleus. If it were known how well the nucleus reflects sunlight, the diameter could easily be calculated by its brightness when it is at great distances from the sun and therefore inactive. Unfortunately it is not known. If the nucleus is a good reflector like clean snow, its diameter is about a kilometer. If it is a poor reflector like the moon, its diameter might be four to six kilometers or even more.

Fortunately one can estimate the diameter through another approach: by placing limits on the rate at which the comet is losing mass. Knowing the velocity of ejection and the accelerations thereby produced one can calculate the total mass. In 1973 two French investigators, J. L. Bertaux and Jacques E. Blamont, estimated the mass being lost by Comet Encke with the aid of instruments in a satellite, which measured the ultraviolet radiation emitted by hydrogen atoms being blown away from the comet. If one assumes that all the hydrogen comes from water molecules dissociated by sunlight, one can arrive at a lower limit for the total mass lost by the comet with each revolution around the sun. It is 6.5×10^8 kilograms, or 650,000 metric tons.

The minimum reasonable value for

TIPPING FORCE ON A COMET NUCLEUS arises if the nucleus is slightly oblate, which is likely, rather than spherical. A rigid or nearly rigid rotating body will rotate stably about its shortest axis. Gases that are being volatilized from the comet's surface by solar heating will exert a reactive force perpendicular to the surface. In this illustration the average force passes below the center of the comet, which tends to push the spin axis to the left. Because the comet is spinning, however, the actual movement of the pole is 90 degrees away from a plane that is defined by the force and the spin axis. In this case the movement is into the plane of the page.

the bulk density of a comet nucleus is one gram per cubic centimeter, the density of water. One can then use the known velocity of the jet action to calculate that the diameter of Comet Encke is greater than 1.2 kilometers. Similarly, one can use estimates of the loss of dust made by Sekanina and Hans E. Shuster to calculate that the diameter is probably less than 2.6 kilometers. Hence the best estimate we can make at present for the diameter of the nucleus of Comet Encke is two kilometers. If that figure is approximately correct, the nucleus reflects 25 percent of the incident sunlight and is a dirty gray, about as one might expect.

The loss of, say, two million tons of ice per revolution corresponds to a mass loss of only one part in 2,000, or an average radius loss of only 16 centimeters per revolution, or 13 meters integrated according to Comet Encke's changing brightness over the 59 revolutions since the comet was discovered in 1786. The shape of the nucleus would change very little in that time, although it is perhaps being flattened a bit because of the different rates of mass loss on its two hemispheres. A mass loss of much less than one part in 1,000 per period is all that is needed to maintain Comet Encke's observed activity, its nongravitational motion and its spin-axis gyration.

Sekanina and I have attempted to determine the period of rotation for Comet Encke by a method I devised three years ago. Many comets, including Comet Encke, develop areas on their surface that actively eject gas when the rotation of the comet exposes those areas to the sun. The result of the periodic ejection of gas is a series of concentric or nearly concentric halos, some of which can be measured on photographic images of the coma. Comet Donati, the great comet of 1858, is the most conspicuous example. From measurements of the diameter of such halos and a knowledge of the expansion velocity as a function of distance from the sun, one can calculate "zero dates": times when the areas that eject gas are turned toward the sun and become active. The zero dates should be separated by multiples of the period of rotation. Comet Donati ran like clockwork with a period of 4.6 hours for three weeks. Usually, however, the observations are so scattered and the uncertainties in the zero dates are so great that several solutions for the period will fit the observations. From four apparitions of Comet Encke in the 19th century and five in this century we find a probable mean value of 6½ hours for the rotation period of the comet's nucleus. We consider this result indicative but not definitive. There is also some evidence for a shortening of the rotation period of possibly 20 to 40 minutes per century.

With a period of 6½ hours and a di-

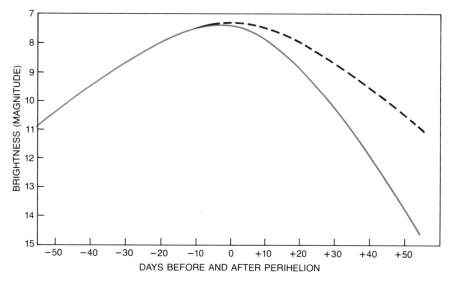

LIGHT CURVE FOR COMET ENCKE (*color*) exhibits a peculiar asymmetry. After perihelion the comet's brightness fades much faster than it had increased before perihelion. The curve, corrected to a standard earth distance, is based on observations of 11 perihelion passages between 1937 and 1974. The broken black curve is the postperihelion luminosity expected if comet's fading mirrored its brightening. One magnitude equals a factor of 2.51 in brightness.

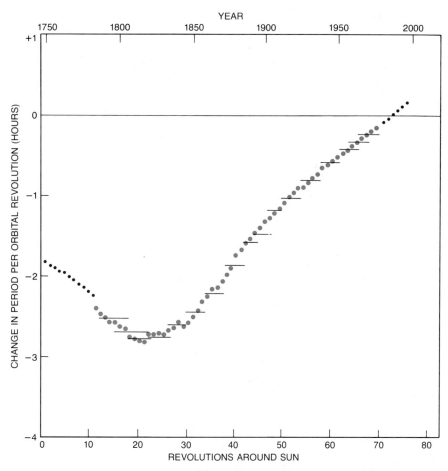

CHANGES IN THE PERIOD OF COMET ENCKE can be attributed to small cumulative changes in the comet's kinetic energy produced by the volatilization of gases from its surface. The magnitude and direction of the jet forces acting on the comet vary enormously with the comet's distance from the sun and with the precession of the comet's spin axis. The author and Sekanina devised separate computer programs to integrate the effect of the jet forces acting on Comet Encke from 1786 to 1977. The colored dots are the computed changes in the comet's period for each revolution. Black dots are extrapolations. Black bars are the actual changes in the comet's period averaged over four or five apparitions for each bar. Sekanina was assisted in his calculations by Brian G. Marsden of the Smithsonian Astrophysical Observatory.

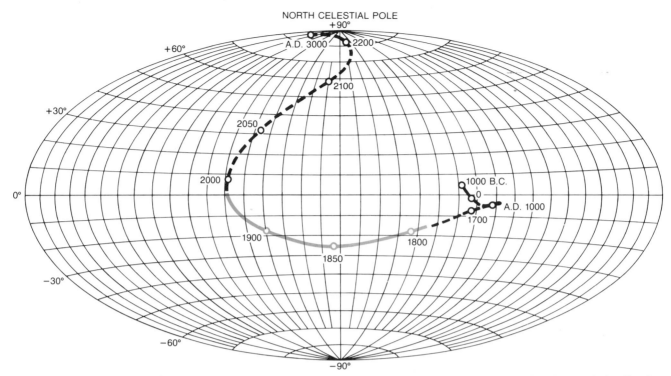

NORTH CELESTIAL POLE

POLAR MOTION OF THE SPIN AXIS of Comet Encke shows wide excursions when it is plotted on the celestial sphere. Solid curve, based on observations and computer studies, shows pointing direction of spin axis from 1786 to 1977. The broken curves are extrapolations.

ameter of two kilometers the required oblateness of the nucleus of Comet Encke is only about .03. The meaning of this figure is that the diameter of the nucleus at the spin axis need be only 60 meters shorter than the diameter at the equator in order to give rise to the observed gyrations of the spin axis. Actually the nucleus is probably covered with pits, craters, mounds and pedestals, so that what is being discussed is only a

systematic trend in a very rough-hewn body that averages out to be spherical within 3 percent.

One may ask: If comets are only little dirty snowballs, why is their study of any importance with respect to the study of other astronomical bodies? The answer is straightforward. Comets clearly represent the most primitive bodies left over from the making of the sun and the planets. The interstellar material that

formed comets may never have been heated significantly. Comets or bodies like them were the building material of the great outer planets Uranus and Neptune. Hence the study of comets can be expected to solve some of the puzzles about the formation of the earth and the rest of the solar system.

Another reason for studying comets is their possible role in making life on the earth possible. The conjecture is,

CONCENTRIC HALOS are sometimes observed in the coma, or diffuse envelope, of a comet. They are evidently produced when the rotating nucleus of the comet presents a particularly active surface to the sun. The gas ejected from the active region forms halos at intervals that coincide with the comet's day. Perhaps the most remarkable halos were those of Comet Donati. These two drawings are based on visual observations on successive days by G. P. Bond of the Harvard College Observatory in the year of the comet's discovery, 1858. The spacing of the halos indicates that the comet was rotating once every 4.6 hours. Halos were generated like clockwork for three weeks.

I would grant, controversial. The evidence is nonetheless clear that when the earth was young, it was too hot to hold the primordial atmosphere supplied by the coalescence of the nebula that gave rise to the sun and the planets. Geologists have argued that the outgassing of volatile substances from the earth's interior was sufficient to supply a second atmosphere as the earth cooled. On the other hand, it is known that the earth, like the other inner planets and the earth's moon, was bombarded during the first half-billion years of its existence by a huge number of smaller bodies, including many whose composition must have been cometlike. Comets could therefore have supplied a not inconsiderable fraction of the water, nitrogen, oxygen and carbon from which life on the earth developed. Moreover, interstellar dust and comets are believed to contain a variety of organic compounds that could provide a good start toward the evolution of living organisms. Fred Hoyle has speculated that life itself originated in "little warm pools" in comets. One need not go that far to hope money will be found to support a space mission to examine a comet at close range.

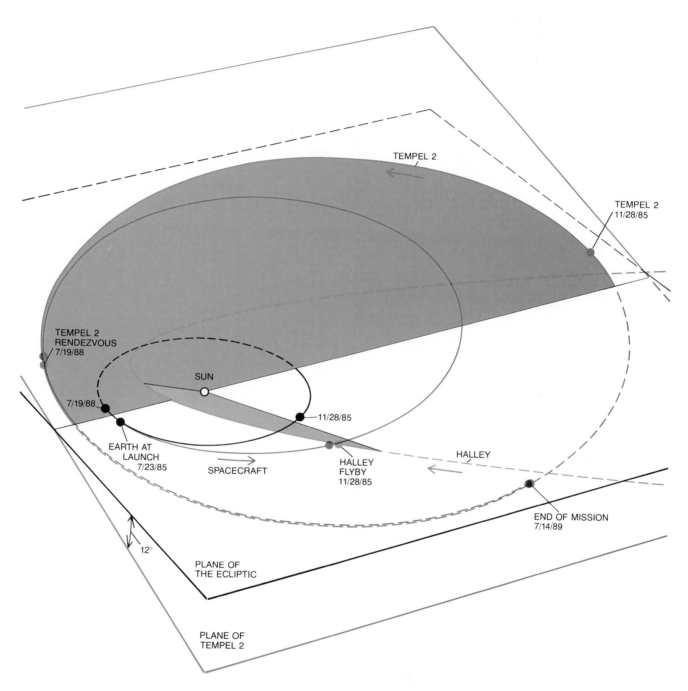

SPACE MISSION TO TWO COMETS will be attempted in the near future if Congress approves a pending proposal of the National Aeronautics and Space Administration. The key to the mission is an advanced propulsion system in which electricity from solar cells will power a cluster of ion-drive rocket engines. After the spacecraft has been lifted into earth orbit by the space shuttle and boosted to escape velocity by a solid-fuel rocket the ion drive will take over and provide continuous thrust for the rest of the three-year mission. If the spacecraft is approved, it will be launched in July, 1985. Four months later, as the craft passes between the sun and Halley's comet, it will drop **off a probe aimed directly at the comet's nucleus. The spacecraft will continue on its course and rendezvous with Comet Tempel 2 in July, 1988, observing the comet from a distance of less than 2,000 kilometers as the comet passes through perihelion. When the comet has proceeded farther along its orbit and has settled into a quiescent state, the spacecraft will approach it close enough to be captured by it and will circle its nucleus at a distance of 10 kilometers. After a year a landing on the comet nucleus will be attempted. This illustration of the mission was prepared with the help of Rolf C. Hastrup of the Jet Propulsion Laboratory of the California Institute of Technology.**

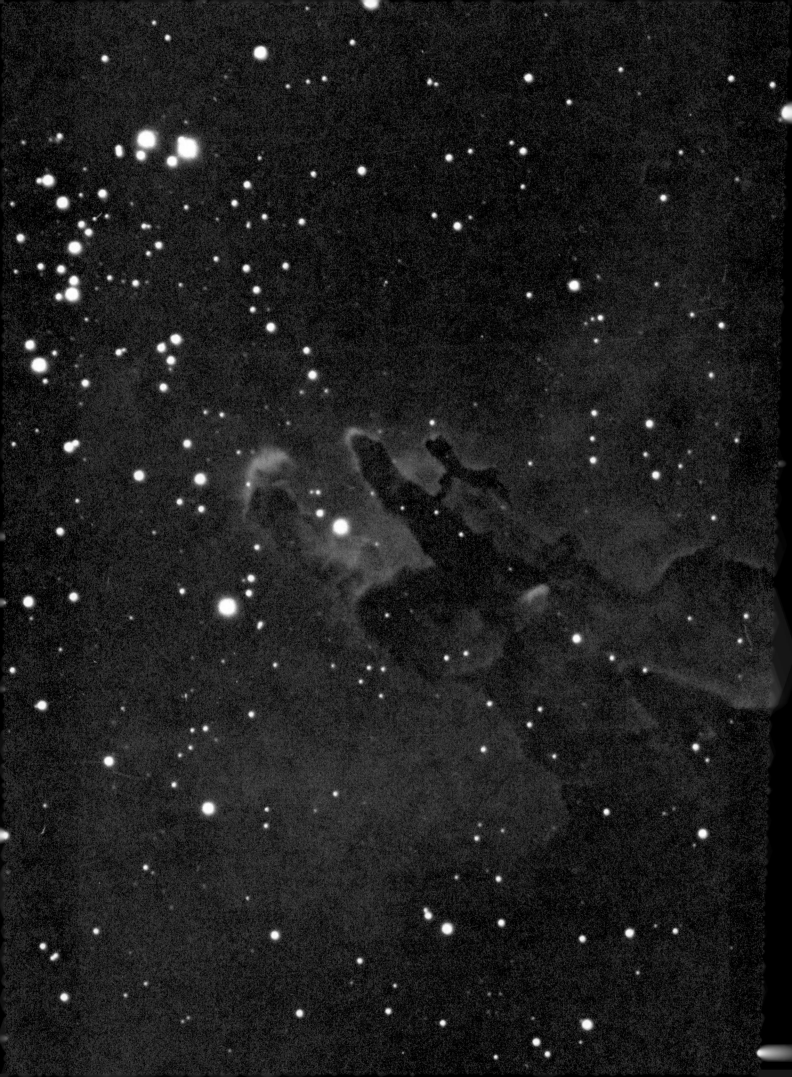

The Origin and Evolution of the Solar System

by A. G. W. Cameron
September 1975

It is generally agreed that some 4.6 billion years ago the sun and the planets formed out of a rotating disk of gas and dust. Exactly how they did so remains a lively topic of investigation

A great cloud of gas and dust contracted through interstellar space 4.6 billion years ago, far out along one of the curved arms of our spiral galaxy. The cloud collapsed and spun more rapidly, forming a disk. At some stage a body collected at the center of the disk that was so massive, dense and hot that its nuclear fuel ignited and it became a star: the sun. At some stage the surrounding dust particles accreted to form planets bound in orbit around the sun and satellites bound in orbit around some of the planets.

So goes—in very broad outline—the nebular hypothesis of the origin of the solar system. Its central idea was proposed more than 300 years ago. It sounds simple enough, and it makes intuitive sense to the layman; indeed, some version of it is accepted by most astronomers today. And yet beyond the broad outlines there is no consensus among students of the origin and evolution of the solar system. We still have no generally accepted theory to explain how the primitive solar nebula formed, how and when the sun began to shine and how and when the planets coalesced out of swirling dust.

It was René Descartes who first proposed (in 1644) the concept of a primitive solar nebula: a rotating disk of gas and dust out of which the planets and their satellites are made. A century later (in 1745) Georges Louis Leclerc de Buffon put forward a second theory: that a massive body (he suggested a comet) came close to the sun and ripped out of it the material that constituted the planets and their satellites. In the two centuries after Buffon the many theories that were propounded tended to follow in the tradition of either Descartes's monistic view or Buffon's dualistic one; the balance of favor swung back and forth between them. The most significant early monistic theories were those of Immanuel Kant and Pierre Simon de Laplace, who elaborated on Descartes's original idea by explaining how the cloud of gas and dust, shrinking to form the sun, would have spun faster and faster because of the conservation of angular momentum: a decrease in the radius of a rotating mass must be balanced by an increase in its rotational speed. Laplace suggested that a series of rings were shed, from whose dust the planets and satellites were formed. At the end of the 19th century dissatisfaction with the ability of the nebular hypothesis to explain the accretion of matter into the planets brought dualistic theories back into favor. Today they have been generally abandoned; it seems clear that most of the material that might have been drawn out of the sun by, say, the approach of another star would have fallen back into the sun or dispersed in space before any solid condensates could coagulate into planets.

A major reason for the wide range of early theories of the origin of the solar system was the lack of observational data—of facts to be explained by a theory. The history of the earth's first few hundreds of millions of years is missing from the geological record, which could therefore offer no clues to the environment in which this sample of a planet was born, and the limited capabilities of telescopes restricted the astronomical data. The early theories were devised to explain only a few observations: the spacing of the planetary orbits increased in a regular way (in accordance with what is known as Bode's law); planetary orbital motions and spins tended to have the same direction of rotation; the sun accounted for only a small fraction of the total angular momentum of the solar system, even though it accounted for the greatest fraction by far of the total mass of the system. These few facts provided few constraints on theory, and so the theories proliferated.

In just the past three decades the situation has changed dramatically. We have a vast amount of new information that imposes additional and powerful constraints on any theory. The new knowledge stems notably from new research on meteorites and from the data returned to the earth by spacecraft dispatched to other bodies in the solar system.

The meteorites are samples of primitive solar-system material. They are evi-

NEW STARS ARE BORN, as the sun may have been born, in gaseous emission nebulas: diffuse, dusty clouds of hot interstellar gas. The photograph on the opposite page, made with the Mayall four-meter reflecting telescope at the Kitt Peak National Observatory in Arizona, shows the nebula designated M16 or NGC 6611, called the Eagle Nebula, in the local arm of our galaxy. Within the nebula clouds of gas have condensed relatively recently to form bright blue-white stars, and other such clouds are still condensing. Ultraviolet radiation from the hot new stars ionizes hydrogen atoms in the remaining gas, giving rise to free electrons and protons. When high-energy electrons recombine with protons, light is emitted at the hydrogen-alpha wavelength of the spectrum: red light that illuminates the cloud and silhouettes dense, dusty, cooler regions of the nebula where the light does not penetrate.

a

b

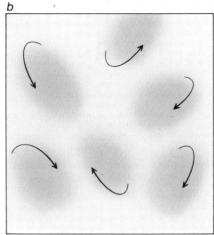

c

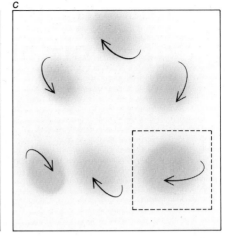

GALAXIES FORMED in the thin, expanding primordial gas (mostly hydrogen, with some helium) when regions of somewhat greater density (*a*) contracted gravitationally to form protogal- axies (*b*), rotating because of the net effect of gas eddies within them. The protogalaxies continued to contract gravitationally, and then to rotate faster (*c*). One of them (*rectangle*) was our own.

dently fragments of rather small bodies that have collided and broken up, sending many of their pieces into new orbits that ultimately intersect the earth. They bring to us, trapped in their interior, samples of the gases of the solar nebula. The details of their mineralogy provide clues to the temperatures and pressures in the nebula at the time its individual grains were last exposed to chemical reaction with its gases. From the relative amounts of the products of radioactive decay that remain trapped in the interior of the meteorites we learn how long ago the original elements that gave rise to certain radioactive isotopes were assembled to form the meteorites' parent bodies.

One of the primary scientific goals of the space-probe program was to advance understanding of the origin of the solar system, and the program has already borne fruit. Measurements made by spacecraft have refined our knowledge of planetary masses and radii, from which we derive accurate mean densities of the planets and clues to their internal composition. By observing how the gravitational potentials of a planet differ from those of a perfectly uniform sphere we derive constraints on the degree to which the density can vary in different parts of the planet's interior. Determining whether or not a planet has an intrinsic magnetic field tells us something about the planet's internal dynamics. Spacecraft data on the composition of a planetary atmosphere reveal something about the gases that once were incorporated in the planet and about chemical interactions between the atmosphere and the planet's surface. Examining the incredibly detailed images of solid plane-

tary surfaces that have been sent back by spacecraft cameras, we can see how volcanic and other geological processes have operated on other planets. The density of craters tells us about the terminal stages of the planet's accretion and about the numbers of smaller bodies that have wandered through the solar system.

Still other constraints come from the general advances in astrophysics that have marked the past three decades. We now know that our galaxy as a whole is between two and three times older than the solar system; we therefore have good reason to believe that the conditions we see today in the galaxy are not very different from those at the time the solar system was formed. We see regions in our galaxy in which stars have been formed in the recent past and are probably still being formed today; that gives us important information if we believe the sun and the solar nebula formed as parts of the same general process. We have learned much about the birth and death of stars and how elements originate in nuclear reactions within exploding stars and are formed into tiny grains of interstellar dust, and about how those grains concentrate in the dark patches in the sky that blot out the light coming to us from distant stars. Those grains of dust and the interstellar gases that accompany them were the raw material of the solar nebula. Let me now try to weave the many threads of information into a coherent picture of the solar system's formation.

Galaxies form when gas—mostly hydrogen—collapses out of intergalactic space. Many billions of years before

the origin of the solar system our galaxy began to take shape in that way. Out of the collapsing gas a first generation of stars was born—stars that still remain spherically distributed around the center of the galaxy, a reminder of its original roughly spherical shape. After those first stars were formed the residual gas, because of its intrinsic angular momentum, settled into the thin disk that is a characteristic feature of all spiral galaxies, and further generations of stars formed from the gas in the disk. The more massive of them evolved quickly, forming heavy elements that were ejected into the interstellar gas. Some of the heavy elements condensed into tiny grains: the interstellar dust. When enough stars had formed in the central plane of the galaxy, an instability de-

a

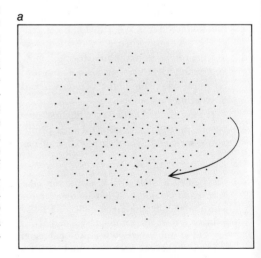

OUR GALAXY EVOLVED as dense lumps of gas contracted within the protogalaxy to form a first generation of stars (*a*). In time the residual gas settled into a disk in the

veloped in their motions that allowed them to cluster together temporarily, forming the spiral arms.

Such arms represent local enhancements of star-population density in the disk; the arms are continuing features that rotate around the center of the galaxy, but the material that constitutes them keeps changing: individual stars spend only about half of their time in one arm before moving on to the next one. Like the stars, the interstellar gas and dust spend about as much time in an arm as they do flowing through the larger spaces between successive arms; the result is that the density of gas and dust is considerably enhanced in a spiral arm. We know from studying galaxies other than our own that it is in these high-density spiral arms that the new stars of spiral galaxies are formed.

Pressure differences arise within the gas, perhaps as the result of a supernova explosion; the gas flows away from regions of higher pressure, but in moving it may tend to pile up somewhere else. Clouds of high-density un-ionized gas accumulate, which typically have a mass of from several hundred to several thousand times the mass of the sun. Gravitational forces tend to pull such a cloud into a more compact configuration. Contraction is opposed, however, by the internal pressure of the gas in the cloud, which tends to make the cloud expand; ordinarily the internal pressure is much stronger than the gravitation and the cloud is in no danger of collapsing. Sometimes, however, a sudden fluctuation in pressure—from a nearby violent event such as a supernova explosion, the formation of a massive star or a large re-

arrangement of the interstellar magnetic field—may compress a cloud to a density much higher than normal. Under such conditions, which are quite rare, gravity may win out over internal pressure, so that the cloud begins to collapse to form stars. As the cloud collapses, its interstellar grains shield its interior against the heating effect of radiation from the stars outside. The temperature of the cloud falls, and the internal pressure becomes less effective. The collapsing cloud breaks into fragments and the fragments break into smaller fragments. When one small fragment eventually completes its collapse, it will have formed into a flattened disk, cool at the edges and very hot at the center: a primitive solar nebula.

What was the nature of the solar nebula and how did it evolve? When did the sun form? Why are there planets? How did they take shape? Quite different pictures of the structure of the primitive solar nebula and of its evolution result from different estimates of its size. Such estimates have usually been arrived at by reasoning backward in time from the present masses of the planets. Let me reproduce such an argument.

In a very general way one can divide the materials of the planets into three classes depending on their volatility: rocky, icy and gaseous. The major constituents of rocky materials are iron and oxides and silicates of magnesium and other metals, notably aluminum and calcium. All these materials would be in solid form at pressures characteristic of the primitive solar nebula and at tem-

peratures in the range from 1,000 to 1,800 degrees Kelvin (degrees Celsius above absolute zero). The four inner planets and the earth's moon (and at least two of the major satellites of Jupiter) appear to be basically rocky. The rocky solids represent about .44 percent by mass of the material out of which the sun formed. The present mass of a rocky planet, then, represents about .44 percent of its share of the primitive solar nebula; the remainder of that share is "missing" because it was too volatile to have been incorporated in the planet.

At a temperature below 160 degrees K. the water in the nebula would be in the form of ice. Ammonia and methane form solids only at a somewhat lower temperature. The ices constitute 1.4 percent by mass of the material out of which the sun formed. Rock-ice mixtures account for most of the mass of Uranus and Neptune and some of the mass of Saturn and Jupiter (and for the bulk of the mass of most of the satellites of the outer planets and the comets). Arguing as for the inner planets, one can assume that the rock and ice now present in such bodies represent 1.4 plus .44 percent, or 1.84 percent, of those bodies' original share of the nebula.

At any temperature likely to have been attained in the primitive solar nebula the very volatile elements—hydrogen and the noble gases such as helium and neon—would remain in the gaseous state. Such gases are incorporated in bodies within the solar system only to the extent that they have been held in planetary atmospheres by gravity and, in the case of hydrogen, held in chemical compounds such as water. (A tiny amount

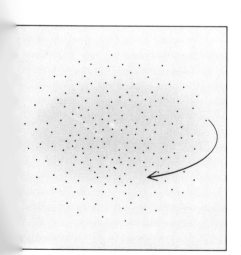

c

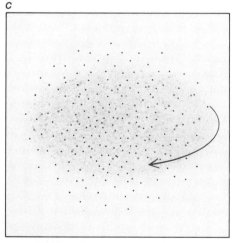

d

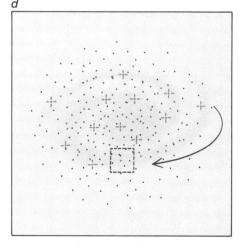

plane of the galaxy's rotation under the combined influence of gravity and centrifugal force (b). Further generations of stars (color) formed within the disk (c); some stars, evolving rapidly, produced heavier elements through nuclear fusion and ejected them into the

disk, where some elements condensed into solid interstellar grains. Instabilities in the motions of gas and stars led to density enhancements that we see as the spiral arms of the galaxy (d). The area in the rectangle is enlarged in the first drawing on the next page.

of helium comes from the decay of radioactive elements.) Morris Podolak and I recently analyzed the structure of the outer planets. We determined that hydrogen and helium constitute about 15 percent of Uranus' mass, about 25 percent of Neptune's, about two-thirds of Saturn's and about four-fifths of Jupiter's. In these planets it is necessary to allow for the gaseous components in order to establish the present rock-ice mass.

By thus establishing the rock and the rock-ice masses of the planets and augmenting those masses for the missing constituents that were too volatile to condense it is possible to estimate a minimum mass for the primitive solar nebula: a mass sufficient to account for the formation of the planets. That minimum mass is about 3 percent of the mass of the sun [see *illustration on opposite page*]. (Older estimates arrived at a much smaller mass—less than 1 percent

of the sun's—because they did not allow for enough rock and ice in Jupiter and Saturn.)

The 3 percent figure is definitely a minimum. It assumes that the planets were completely efficient in collecting from the solar nebula all the material that was in condensed form in each planet's orbit in the nebula. For two kinds of solid, however, that collection process might have been quite inefficient. Consider first the tiny unconsolidated grains of interstellar dust, perhaps a micrometer (a thousandth of a millimeter) in diameter, that were not vaporized as the gas-cloud fragment collapsed. The thickness of the nebular disk must have been at least one astronomical unit (the mean distance between the earth and the sun). That dimension is very large compared with the dimensions of any of the planets, which consolidated approximately in the central plane of the disk. Gas-drag effects

would prevent large quantities of these small grains from settling through the nebular gas toward the central plane at a significant rate; if much of the gas was instead dissipated inward to form the sun, the grains would have accompanied the gas and could never have become incorporated in the planets.

Larger bodies (centimeters or meters in diameter), on the other hand, would fall rapidly through the gas toward the midplane but might nevertheless not end up in planets. As a result of a difference between the centrifugal forces that act on the solid bodies and those that act on the gas, the solids would rotate around the central spin axis of the nebula more rapidly than the accompanying gas. They would therefore move through the gas with a relative velocity as high as several hundred miles an hour; a head wind of that speed would tend to slow them down so that they would spiral rather quickly through the gas toward

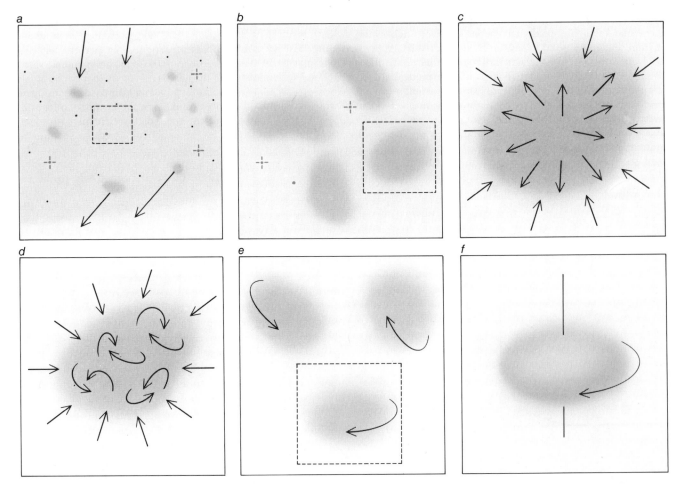

SOLAR SYSTEM EVOLVED in a spiral arm about two-thirds of the way out from the center of the galaxy. Stars, gas and dust grains move through the arm, and new stars are born there; massive, short-lived stars outline the arms (*a*). A supernova explosion or the birth of massive stars creates instabilities that concentrate high-density clouds of gas (*b*). Gravitational forces contract the cloud, but the cloud's internal pressure opposes contraction (*c*). If the cloud has enough mass, gravity dominates (*d*) and the cloud collapses. The collapse generates strong gas eddies (*curved arrows*) and breaks the cloud into fragments (*e*); each fragment has a net rotation derived from its major eddies. One of these fragments spins faster and its gas settles into a disk that was the primitive solar nebula (*f*).

the central spin axis and thus be lost to the region of planet formation. For these two reasons the mass of the primitive solar nebula may have been considerably larger than 3 percent of the sun's mass.

Many of the solar-system theories constructed over the past three decades have involved some version of a minimum-mass solar nebula. The concept has a major flaw, however. It assumes that the sun itself was formed directly during the process of collapse and that the primitive solar nebula was marshaled independently around the sun. The trouble is that simple estimates of the amount of angular momentum that must have been contained in the collapsing cloud fragment indicate that it would have been impossible for almost all of the fragment simply to collapse directly to form the sun, leaving a small fringe of nebula to constitute the planets. Such estimates require instead that the nebula's mass be spread out over several tens of astronomical units. The solar nebula itself must have contained substantially more than one solar mass—and probably about two solar masses—of material, with no sun originally present at the central spin axis. Let me first explain the source of the large amount of angular momentum and then show why it indicates that there was not a minimum solar nebula but a massive one.

The strong fluctuations in pressure that led to the rapid compression of the original interstellar cloud and thus brought it to the threshold of gravitational collapse must have stirred the cloud's gases into violent turbulence. Large-scale shearing motions developed —eddies superimposed on eddies, in a wide range of sizes and in many planes and directions. When any one fragment became isolated from such a turbulent cloud, it had a net tendency to spin, derived from the motions of the largest eddies it happened to contain. A fragment's mass, its rate of rotation and its radius combine to endow it with a certain amount of angular momentum, and that momentum must be conserved; as the fragment contracted, it spun faster. The sun turns very slowly, however; in spite of its great mass it accounts for only 2 percent of the solar system's angular momentum. Most of the original angular momentum of the vast quantities of gas that moved in to form the sun must have been transported outward; a considerable part of the original nebula must therefore have remained at

PLANET	PRESENT MASS (PERCENT OF SUN'S)	AUGMENTED MASS (PERCENT OF SUN'S)
MERCURY	.000017	.004
VENUS	.000245	.056
EARTH	.000304	.07
MARS	.000032	.007
JUPITER	.09547	1.5
SATURN	.02859	.77
URANUS	.00436	.27
NEPTUNE	.00524	.27
PLUTO	.00025 (?)	.06 (?)
TOTAL (MINIMUM MASS OF SOLAR NEBULA)		3.0

MINIMUM MASS OF SOLAR NEBULA is estimated by adding up the amount of solar material that must have been present (*column at right*) to account for the present mass (*middle column*) of each planet. Solar-nebula mass thus estimated is 3 percent of mass of sun.

great distances from the sun to take up that angular momentum.

An additional reason for postulating a massive solar nebula is the observation that young stars tend to lose mass at a prodigious rate early in their lifetime; the loss comes as they pass through what is called their T Tauri stage, which I shall discuss in a bit more detail below. The combination of the mass that remained in the solar nebula and never became part of the sun and the mass that was once in the sun but was lost in the early sun's T Tauri stage could easily have amounted to as much as one solar mass.

As a result of this kind of reasoning—in effect arguing forward from what is known of the principles of star formation rather than backward from the masses of the present planets—Milton R. Pine and I constructed some numerical models of the massive solar nebula. The models extended out to a radial distance of about 100 astronomical units and contained two solar masses of material. In a typical model the temperature was about 3,000 degrees K. near the spin axis and decreased to a few hundred degrees in the region of planet formation. Such temperatures are considerably higher than the temperatures that characterized the collapse of the original interstellar cloud; they develop in the later stages of compression of the gas, once its density becomes high enough so that its own cooling radiation can no longer escape easily. The escape of this radiation is impeded, however, only during the rapid final stages of the collapse; once the gas stops contracting— once the primitive solar nebula is formed

—the radiation can escape relatively quickly, so that in the region of planet formation the nebula will lose most of its heat energy in only a few hundred or a few thousand years.

Such a short cooling time (short compared with the time required to form a sun and planets) presents a difficulty for the massive-nebula model. As the nebula cools it will flatten into a thinner disk, and thin disks have been shown to be dynamically unstable: they tend to deform into a barlike configuration. (Such a deformation might well be the mechanism by which close pairs of double stars are formed, but that evidently did not happen in the solar system.)

There is another time-scale problem for the massive-nebula model. An important process for transporting angular momentum away from the central spin axis so that gas can shrink toward that axis is probably a system of fast meridional currents: gas currents that flow in a plane parallel to the spin axis and at a right angle to the central plane of the nebula. Pine and I estimated that the characteristic time for the outward transport of the angular momentum shed by the inner parts of the primitive solar nebula would be only a few thousand years. John Stewart of the Max Planck Institute for Physics and Astrophysics in Munich has shown that gas turbulence must play an important role in a primitive solar nebula and may cause an even more rapid outward transport of angular momentum.

Both the time for cooling and the time for angular-momentum transport seem too short compared with the time required for the accretion of the solar nebula. After a fragment separates from the interstellar cloud its central region is likely to be denser, and will collapse more rapidly, than the remainder of the fragment. A small solar nebula will therefore be formed at first when the central region ceases to collapse; that small nebula will grow by accretion of the remainder of the infalling fragment

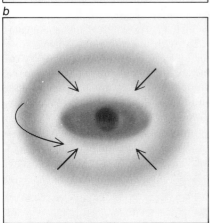

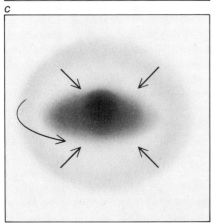

ACCRETION MODEL of the primitive solar nebula assumes that a central region of the cloud fragment collapses faster than the rest (a). It forms a small solar nebula: a central mass that is not yet the sun, surrounded by a disk of gas and dust grains, with more gas and dust concentrated around the periphery (b). The small nebula then grows by accretion over a long period of time (c).

over a period of time—probably between 10,000 and 100,000 years, a lot longer than the cooling and angular-momentum transport times we estimated.

These considerations have led me to a new picture of the primitive solar nebula that I am currently trying to define in detail. It is necessary to construct not a single model but an evolutionary sequence of models, beginning with a small solar nebula that grows through accretion over a time interval of perhaps 30,000 years. In this case the time for the redistribution of angular momentum remains short compared with the accretion time, so that much of the mass flows inward to form the sun not at the beginning of the accretion period but throughout that period; the mass of the nebula out in the region of planet formation remains a relatively small fraction of a solar mass throughout the period. As for cooling, the accreting gas is suddenly decelerated when it hits the surface of the solar nebula; the energy of its infall is converted into heat that is radiated from the surface. In the later stages of accretion that process keeps the surface layers in the region of planet formation at a temperature of perhaps a few hundred degrees; the temperature in the interior would be somewhat higher. And meanwhile the steady flow of mass toward the central spin axis diminishes dynamic instabilities within the nebula.

The two pictures of the primitive solar nebula, one derived from the masses of the planets and the other from the principles of star formation, thus seem to be converging to form an intermediate model of the initial solar nebula. In that model somewhat more than one solar mass has collected toward the spin axis but is not yet recognizable as the sun. It is surrounded by a disk of gas and dust amounting to perhaps a tenth of a solar mass. Farther out, beyond the region of planet formation, considerable additional amounts of mass are still falling toward the solar nebula.

The planets were created by the accumulation of interstellar grains and, in the case of the outer planets, the subsequent attraction and adherence of gases. The buildup of solid matter would have begun, I have recently calculated, in the collapsing gas cloud. Turbulent gas eddies would have accelerated the interstellar grains until they had large enough relative motions to begin to collide with one another. Having been formed out of material in stars and then ejected into interstellar space, where ices and other volatile constituents condensed on their

surface, the grains probably had a rather fluffy structure. It would not be surprising if such particles stuck to one another when they collided, forming clumps. As time passed the clumps of grains would collide with one another, sometimes amalgamating into larger clumps and sometimes breaking up into smaller ones. By the time the solar nebula had formed, many clumps were likely to have grown to a diameter measured in millimeters or centimeters.

Clumps of that size could settle through the gas toward the midplane of the nebula in tens or hundreds of years. Since their settling rate would vary with size there would be further collisions, increasing the size of the clumps and accelerating their fall toward the midplane. At that point, however, unless they were somehow able to grow substantially larger they would rapidly be lost to the inner solar nebula as a result of the gas-drag effect I mentioned above.

A critical process in planet formation may therefore be a mechanism recently proposed by Peter Goldreich of the California Institute of Technology and William R. Ward of Harvard University, which would give rise to those larger bodies. They showed that if there is a thin layer of condensed solids at the midplane of the nebula, with very little relative velocity among the particles, then a powerful gravitational instability mechanism will break up the thin sheet into bodies with diameters in the range of the diameters of asteroids: kilometers or tens of kilometers. The instability mechanism gradually operates over larger distances, attracting the asteroid-size bodies into loosely bound gravitating clusters of hundreds or thousands of bodies. The clusters remain unconsolidated because of the large angular momentum contained in their component bodies, which makes them rotate around common gravitating centers. When two clusters approach each other, however, they intermingle; the fluctuating gravitational field in the combined cluster leads to a violent dynamic relaxation of the motions of the bodies, so that many of them coalesce to form cores around which others go into orbit (although some of the bodies would be lost). The clusters interact with one another gravitationally over quite large distances; mutual perturbations gradually build up the velocities of the clusters with respect to one another, leading to further collisions that produce ever larger bodies.

The Goldreich-Ward instability mechanism would appear to be a powerful first step in the accumulation of planetary bodies. The subsequent stages in

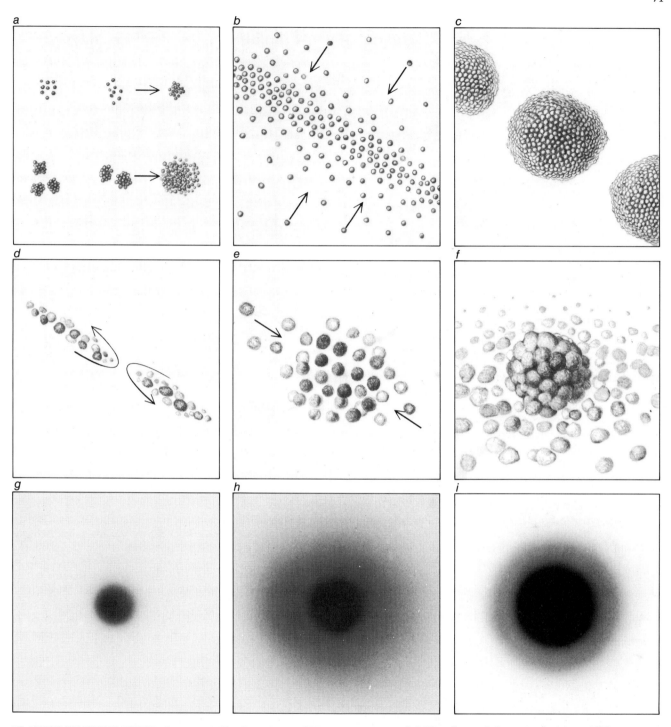

PLANETS BEGIN TO FORM when interstellar dust grains collide and stick to one another, forming ever larger clumps (a). The clumps fall toward the midplane of the nebula (b) and form a diffuse disk there. Gravitational instabilities collect this material into millions of bodies of asteroid size (c), which collect into gravitating clusters (d). When clusters collide and intermingle (e), their gravitational fields relax, and they coagulate into solid cores, perhaps with some bodies going into orbit around the cores (f). Continued accretion and consolidation may create a planet-size body (g). If the core gets larger, it may concentrate gas from the nebula gravitationally (h). A large enough core may make the gas collapse into a dense shell that constitutes most of the planet's mass (i).

the process are still highly speculative and were surely different in different regions of the solar nebula. The interstellar grains whose clumping initiates the accumulation process are those that have not been vaporized by the heat of the nebula; their materials, and therefore the materials of the larger bodies into which they are incorporated, would be different at different distances along the steep temperature gradient: metals, oxides and silicates in the region of the inner planets; similar rocky compounds and water ice farther out; rock, water ice and frozen methane and ammonia still farther out.

In the case of the smaller inner planets the progression to full size may be just a question of successive collisions and amalgamations of rocky bodies. In the case of the outer planets there are other considerations. Fausto Perri and I have recently considered the behavior of the primitive solar nebula as a large planetary core grows within it. As the mass of the core increases, gas in the solar nebula becomes gravitationally concentrated toward the core; with the continued growth of the core the amount of mass in the gas that is concentrated increases even more rapidly than the mass of the core itself. At some point the core reaches a critical size (which depends on the temperature conditions in the surrounding gas) such that the gas becomes hydrodynamically unstable and collapses onto the planetary core.

The major constituents of Jupiter and Saturn are the hydrogen and helium in their atmosphere, and we believe it was through this process of concentration and collapse that these planets acquired most of their mass. Hydrogen and helium account for a smaller fraction of the mass of Uranus and Neptune, probably indicating that their core never grew to the critical size for hydrodynamic collapse; those two planets did, however, grow large enough to retain much of the hydrogen and helium that was gravitationally concentrated toward their core. The inner planets, on the other hand, may be too small ever to have concentrated much of the nebular gas.

When the collapse events took place to form Jupiter and Saturn, local conservation of angular momentum in the gas would cause it to flatten into a disk around the planetary core. As time went on the two planets would sweep up essentially all the gas in their vicinity within the nebula. One can think of them as forming miniature versions of the primitive solar nebula: a central core of condensed rock and ice taking the place of the sun, with the gaseous disk around the core as the analogue of the solar nebula. Both of these large planets have systems of regular satellites incorporating considerable mass, which probably formed from a gaseous disk by processes quite analogous to the formation of the planets in the solar nebula.

Any theory of the origin and evolution of the solar system must account somehow for the comets, its most spectacular but least understood members. Jan Oort of the Leiden Observatory suggested some years ago that the comets inhabit an enormous volume of space centered on the sun, starting well beyond the outer planets and extending to a distance of perhaps 100,000 astronomical units. The total mass of the comets in this vast "Oort cloud" is probably equivalent to between one earth mass and 1,000 earth masses, which would account for between 10^{12} and 10^{15} comets. The comets we see are those few whose orbital elements are perturbed by a passing star in just the right way to send them plunging toward the center of the solar system.

A comet is a "dirty snowball," an aggregate of ice and rocky material, in the model first suggested by Fred L. Whipple of Harvard University. As a comet approaches the sun, gases are vaporized from it, accompanied by dust particles, to form the characteristic coma and tail. Analysis of the tails shows that the molecules that are vaporized are primarily water but also include exotic organic compounds. The dust, some of which comes to rest high in the earth's atmosphere, consists of fluffy clumps of fine-grained rocky material. A comet, in other words, is apparently an assembly of interstellar grains.

The comets must either have been made within the solar nebula and somehow ejected into the Oort cloud or else have been made out in the Oort cloud itself. Oort originally suggested that they were formed near Jupiter and perturbed by Jupiter's gravitational field into very large orbits that were subsequently rounded out by stellar perturbations. That would require the formation of a staggeringly large mass of comets, since many more would have been ejected from the solar system than were retained in the Oort cloud. Moreover, given the temperatures that must have prevailed near Jupiter it seems unlikely that molecules more complex than water would have been in solid form. More complex ices are possible farther out in the nebula, and so Whipple and others have suggested that the comets were formed in the neighborhood of Uranus and Neptune and sent out into the Oort cloud by the gravitational fields of those planets. That proposal, however, meets only one of the objections to the Oort hypothesis.

My own belief is that the comets were probably formed out in the Oort cloud itself. It is true that the collapsing gas of the fragment of cloud that became the primitive solar nebula was never dense enough so far from the center for the interstellar grains to have aggregated into sizable bodies out there. There is another possibility, however. Most of the stars in the galaxy are much less massive than the sun, suggesting that gas-cloud fragmentation sometimes continues at least down to fragments a tenth of a solar mass in size. Fragmentation may have gone further still. Small fragments could have been bound gravitationally to the primitive solar nebula in

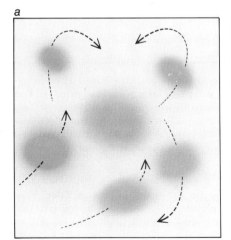
a

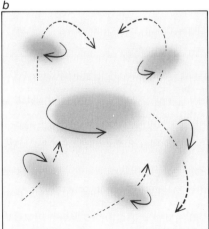
b

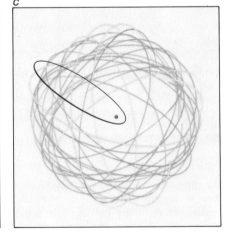
c

COMETS MAY HAVE FORMED from small cloud fragments that once were in orbit around the larger fragment that became the solar nebula (a). The small fragments spun down, like the solar one, to form disks in which comets were accumulated much as planets were (b). Eventually starlight could have evaporated the gases of these "cometary nebulas," leaving the comets in enormous orbits around the sun (c). From time to time a comet's orbit is perturbed by a passing star, and the new orbit brings it close to the sun.

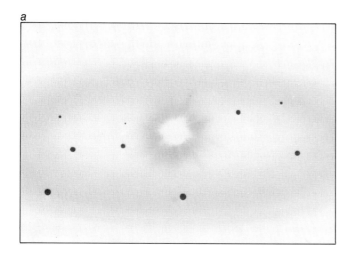
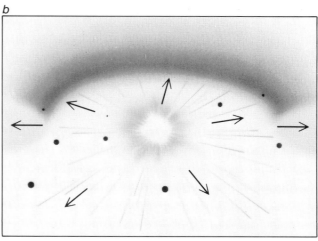

SOLAR SYSTEM WAS CLEANED UP by the "T Tauri wind." When gas contracting toward the center of the nebula reached a sufficient density, the nuclei of hydrogen atoms began to fuse and the sun began to shine (*a*). During its T Tauri phase the sun lost vast quantities of material. That material constituted an intense solar wind that could have blown away the remaining gas (*b*).

orbits traversing the region of the Oort cloud. Such fragments would form fairly large and very cool disks, ideal places for the comets to form. When the disks were ultimately heated by ultraviolet radiation from external stars, the gases would evaporate away and leave the comets in solar orbits [*see illustration on opposite page*].

After the planets had formed, much of the gas of the solar nebula must have remained in orbit around the sun, along with countless small bodies and large amounts of unconsolidated dust. There are only planets and asteroids in orbit now, with very little dust and almost no gas. How was the solar system cleaned up? As I mentioned above, young stars characteristically pass through what is called the T Tauri stage, when they eject matter at a prodigious rate: as much as one solar mass per million years! There is every reason to believe the sun passed through a similar phase, and the fierce "wind" of that ejected mass undoubtedly dissipated the solar nebula by carrying the residual gas off into space. That early solar wind would have stripped the inner planets of the remains of any primitive atmosphere of hydrogen and helium from the primitive solar nebula; the outer planets must have formed early enough to capture their hydrogen and helium before the solar wind began to blow. Moreover, if the accretion of gas from the original cloud fragment was still continuing, it would have been terminated by the wind, with the infalling gases being ejected back into interstellar space.

What determined when the T Tauri wind began? Why did it not blow away the primitive solar nebula long before the sun got so large? The thermonuclear reactions of hydrogen that constitute a stellar furnace are ignited only under extreme conditions of high temperature and high density. Perri and I have recently determined that the temperatures of the primitive solar nebula were so low that compressional heating of the gas could not have ignited the sun at a central density comparable to that of the sun today; the density would have had to be at least 100 times higher than it is now. Only then could the sun have adjusted itself into its present configuration, and only then could the thermonuclear furnace have been ignited and could the intense T Tauri wind begin to blow. An enormous amount of mass must have had to be gathered together to achieve such a density. Given the original temperature of the nebula, in other words, the sun had to reach a large mass before it could be a sun.

Once the sun had begun to shine and the T Tauri wind had blown away the gas, the stage was set for the final cleanup of interplanetary space and the completion of planet formation. The orbits of most of the small bodies in the solar system (other than the observed asteroids in their isolated belt) would have been continually modified by planetary perturbations. Over the course of a few hundred million years most such bodies either would have collided with one of the planets (the surfaces of Mercury, the moon and Mars still show the scars of that terminal bombardment) or would have been ejected from the solar system by a major planet, usually Jupiter.

The tiny dust particles were subjected to forces even stronger than gravitational perturbations: the effects of sunlight. The photons the particles absorb from the sun carry no angular momentum; the photons the particles radiate, however, carry off some of the angular momentum of the particles' orbital motion. The sunlight therefore acts as a resisting medium for the particles, making them spiral in toward the sun. Larger solids, up to a kilometer in diameter, are perturbed by sunlight in a different way. As such a body rotates the temperature of a section of its surface increases as long as it is on the sunlit side but decreases while it is on the dark side. One hemisphere of the body therefore emits considerably more radiation than the other. That gives rise to a preferential thrust that can perturb the orbit of the body either toward the sun or away from it, depending on the body's direction of rotation. Such bodies will eventually come close to one of the planets, whereupon they will be absorbed by collision or be ejected from the solar system. In these several ways the sunlight could have acted as a broom to sweep away much of the smaller debris left over from the formation of the solar system.

The account I have given makes a coherent, if incomplete, story. Many of its details remain highly speculative, however, and much of the story may have to be retold as new data test the present theories. In selecting and weaving together facts, ideas and hypotheses I have necessarily been strongly influenced by my own beliefs. The reader should be aware that others would weave quite different tapestries. These ancient questions are still far from being answered.

9

The Solar Wind

by E. N. Parker
April 1964

*Comet tails and other phenomena indicate that a thin,
hot gas of solar particles flows past the earth at
supersonic speeds. This gas is simply the expanding
corona of the sun*

A swift wind of hydrogen blows continuously through the solar system. Emanating from the sun, it speeds past the earth at 400 kilometers per second (about 900,000 miles per hour) and rushes on past the planets into interstellar space. Like a broom, it sweeps up the gases evaporated from planets and comets, fine particles of meteoritic dust, and even cosmic rays. It is responsible for the outer portions of the Van Allen radiation belts around the earth, for auroras in the earth's atmosphere and for terrestrial magnetic storms. It may even play a part in shaping the general pattern of the earth's weather.

The existence of this solar wind, which had long been suspected, has now been verified by space vehicles. They have measured its velocity and density. And studies of another kind have unraveled the mystery of its origin and given us an understanding of its effects.

The realization of the solar wind's existence came only gradually, over a period of several decades. The first explicit assertion that something besides light was coming to the earth from the sun was made in 1896 by the Norwegian physicist Olaf K. Birkeland. He suggested that the aurora borealis might be caused by electrically charged "corpuscular rays" shot from the sun and "sucked in" by the earth's magnetic field near the poles. He was led to this suggestion by the fact that the aurora looked very much like the electric discharge in the then newly invented tubes generating streams of charged particles ("cathode rays").

Birkeland's idea was taken up by the Norwegian mathematician Carl Størmer, who went on to calculate the paths that streams of charged particles from the sun should follow when they entered the earth's magnetic field. As it happened, his theoretical scheme of looped and spiral paths did look like patterns seen in the aurora, but this resemblance turned out to be a coincidence; nothing else in his theory worked. To this day there is still no complete theory explaining how the solar wind produces the aurora, although some interesting ideas are beginning to develop. The fact remains, however, that Birkeland and Størmer were on the right track and started an important new line of thinking by calling attention to the possibility of charged particles coming from the sun.

Magnetic Storms

Further evidence for the sun's emission of particles came from another phenomenon (and many years later). This had to do with the magnetic storms that are associated with the disruption of radio, telephone and telegraph communication. The storms are evidently caused by fluctuations in the earth's magnetic field. Because they usually came a couple of days after a flare on the surface of the sun, they were at first attributed to a burst of ultraviolet radiation from the flare or some similar cause. Then the British geophysicist Sydney Chapman surmised that corpuscular emissions from the sun offered a more reasonable explanation. In the 1930's he and V. C. A. Ferraro carried out a series of calculations and demonstrated that a cloud of ions ejected from the sun, traveling at 1,000 or 2,000 kilometers per second, would reach the earth in a day or two and ruffle the earth's magnetic field as it passed. Their theoretical picture of such a disturbance of the field so closely resembled the actual fluctuations during a magnetic storm that Chapman's idea was widely accepted.

The third manifestation of the solar corpuscles was noted in the late 1940's. This time they came up in connection with fluctuations in the bombardment of the earth by cosmic rays. Scott E. Forbush of the Carnegie Institution of Washington had discovered that the intensity of cosmic radiation reaching the earth was low during the height of solar activity in the sunspot cycle and often fell abruptly during a magnetic storm. In other words, the more active the sun, the smaller the number of cosmic ray particles impinging on the earth. It was at first supposed that this effect must be due to solar-caused changes in the earth's magnetic field and atmosphere that deflected the cosmic particles away from the earth. But the University of Chicago physicist John A. Simpson, who began to keep track of cosmic ray variations with a neutron monitor he had just developed, soon found that the fluctuations were much greater than had been supposed. They could not be produced merely by changes on the earth; they must reflect a rise and fall of cosmic ray intensity in solar-system space as a whole.

Apparently something in the sun's radiation tended to impede the flow of cosmic rays into the solar system, and this obstruction increased when the sun was particularly active. What might the impeding agent be? The general mechanism is to be found somewhere in the magnetohydrodynamic theory of the Swedish physicist Hannes Alfvén [see "Electricity in Space," by Hannes Alfvén; SCIENTIFIC AMERICAN, May, 1952]. He had pointed out that an ionized gas in motion must carry a magnetic field with it. This being so, it was suggested by Philip Morrison of Cornell University and by others that a stream of charged corpuscles from the sun, carrying a magnetic field, would tend to sweep cosmic ray particles out of the solar system, and the effect would be strongest when the solar radiation was most intense. Such a theory would explain the cosmic ray fluctuations.

At about the same time there emerged a fourth and decisive line of evidence for the corpuscular radiation from the sun. For centuries it has been known that the tails of comets always point away from the sun. No matter where a comet may be in its orbit through the solar system, its head is always toward the sun and its gaseous tail streams away. Why is this so? In modern times the almost universally accepted theory has been that it is the pressure of sunlight, pushing the extremely tenuous matter of the comet, that drives the tail in the opposite direction. But in the 1950's Ludwig F. Biermann of the University of Göttingen showed that the pressure of the sun's light was not nearly sufficient to account for the violence with which a comet's gases are blown away from the head. He suggested in-

stead that the only solar radiation that could account for the pushing away of the comet's tail was a stream of actual particles. He pointed out that such radiation from the sun would also account for the existence of excited, light-emitting ions seen in comet tails [see "The Tails of Comets," by Ludwig F. Biermann and Rhea Lüst; page 39].

Biermann's discovery conveyed something else that had considerable bearing on the question of how this corpuscular radiation from the sun originated. Speculation up to that time had centered on two possibilities: that the corpuscles were sent out in bursts by solar flares (which were known to emit very energetic protons, or hydrogen nuclei) or were projected in beams from sunspots (by some unknown electromagnetic acceleration process). But Biermann's evidence now made it plain that the corpuscular radiation could not be coming merely in bursts or isolated beams. The comet tails showed that the radiation

was blowing continuously in all directions outward from the sun. The comet tails were in effect interplanetary "wind socks" demonstrating the existence of a steadily blowing, space-filling radiation. The streaming of the particles might intensify when the sun became particularly active, but it was present all the time, with or without sunspots or flares.

The Corona and the Wind

So it seemed that the flow of corpuscles must stem from something that went on all the time all over the sun's surface. The sun was continuously shooting a thin hail of projectiles in all directions out into space. By what process could it do such a thing? A suggestion of a possible answer came one afternoon in 1957 when I was visiting Chapman at the laboratories of the High Altitude Observatory in Boulder, Colo., where he was then working.

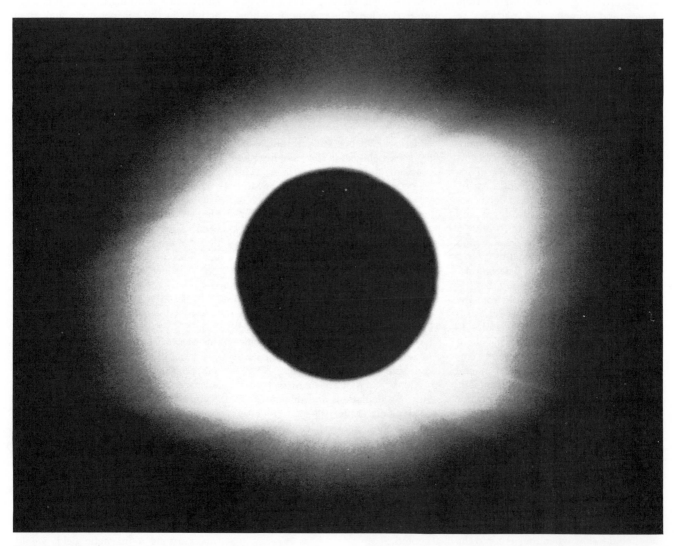

SOLAR CORONA, the source of the solar wind, was in a fairly quiet state during the eclipse of July 20, 1963. The photograph at the left was made at Talkeetna, Alaska, by an expedition from the High Altitude Observatory of Boulder, Colo. That at the

Chapman was studying the sun's corona, from the standpoint of whether or not it might be responsible for heating the outer regions of the earth's atmosphere. Soundings of the upper atmosphere had brought out the curious fact that it got hotter, rather than colder, with increasing altitude. This suggested that the upper air was heated by hot gases in outer space. Chapman suspected that these hot gases might be maintained by the solar corona.

The corona is the tenuous outer atmosphere of the sun. It is very thin indeed: even close to the sun it contains only about 100 million to a billion hydrogen atoms per cubic centimeter, a density only a hundred-billionth that of the air we breathe. The temperature of the corona, however, as measured by the velocity of its atoms, is extremely high: about a million degrees centigrade near the sun. Because of its high temperature the coronal gas is completely ionized and therefore consists of sep-

arate protons and electrons.

Because the corona is so tenuous it is not self-luminous, in spite of its high temperature. It is visible, however, by virtue of the fact that its atoms scatter the light from the sun's luminous photosphere, just as grains of dust in the earth's atmosphere become visible by scattering sunlight. When the brilliant light of the sun itself is dimmed by an eclipse, the white corona can be seen stretching far out from the hidden solar disk. Photographs of the corona show that it extends for millions of miles from the sun, and were it not for interfering haze and light in the sky we could probably see its fainter reaches extending many times farther than that.

Now, Chapman knew from his pioneering theoretical studies of the properties of ionized gases that a tenuous ionized gas at a million degrees must have an extraordinary ability to conduct heat. According to his calculations, the heat flow through ionized gas increases

at a rate almost equal to the fourth power of the increase in temperature. At a million-degree temperature this means a great deal of heat flow. Chapman calculated that, if the corona extended as far as the earth's orbit, its temperature that far from the sun would be about 200,000 degrees, owing to its high conduction of heat. This was a very interesting figure as support for his theory that the corona might heat the earth's upper atmosphere.

But Chapman had made another discovery that impressed me even more as he told me his results in our talk that afternoon. He had gone on to make some calculations to determine if the corona did reach to the earth. For these he used the equation of the barometric law, which states the obvious fact that in an atmosphere the pressure at any given height must be just sufficient to support the weight of the portion of the atmosphere above it. (If it were not, the atmosphere would collapse.) Starting

right was made 115 minutes later by another team from Boulder on Cadillac Mountain in Maine. Careful measurements show motions of solar plumes: either lateral displacements or "virtual" motions due to appearance and disappearance of adjacent plumes.

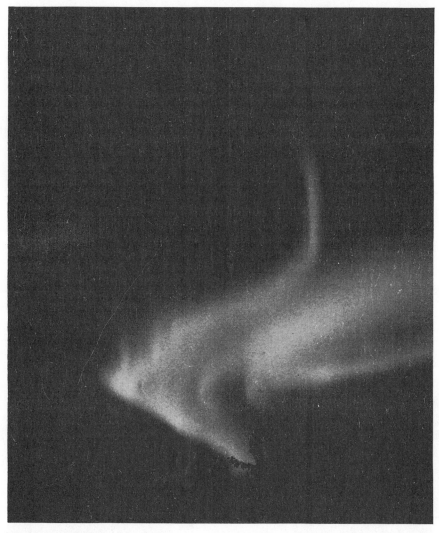

AURORAL DISPLAY, which results from interaction of earth's atmosphere and solar wind, was photographed in Alaska by Victor P. Hessler of the University of Alaska. Lines of force of the earth's field funnel solar particles into the atmosphere at high latitudes.

hydrodynamic equations for the flow of a gas. These nonlinear equations are so complex that it was out of the question at the time to find a general solution covering all possible assumptions; I settled for a simple case that approximately represented what Chapman had calculated, namely, that the temperature of the corona remains high for a distance of several million kilometers from the sun and then drops to a lower figure. This made the mathematics relatively straightforward.

The mathematical solution of the equations produced a result that must be considered surprising in view of the traditional idea that the corona is a static atmosphere. It showed that with increasing distance from the sun the corona tends to expand. At first the expansion is slow, but as the distance increases, the pressure within the corona gradually overcomes the weight of the overlying gas and rapid expansion takes over. At 10 million kilometers (some six million miles) from the sun the corona is expanding at a speed of several hundred kilometers per second—faster than the speed of sound. At that point it must be considered a supersonic wind rather than the sun's atmosphere. It continues to accelerate and reaches velocities several times the speed of sound as it moves out of the sun's gravitational field [*see top illustration on page 83*].

The application of the equations showed that away from the sun the erstwhile corona *must* expand rapidly and become a high-velocity stream. I have called it the "solar wind" because this seems to me now a more accurate description of the phenomenon than the older pictures of a static "atmosphere" or a bullet-like "corpuscular radiation." Biermann's comet tails are truly "solar-wind socks," signaling the direction and strength of the corona's expansion.

The Heat of the Wind

The corona's expansion arises from the fact that its temperature at the sun is of the order of a million degrees. What makes it so hot? We know that the temperature at the sun's photosphere is only about 6,000 degrees, and one would suppose that the nonincan-

from the known density of the corona near the sun (which can be ascertained approximately), he was able to estimate its density at the earth's distance. This turned out to be roughly 100 to 1,000 hydrogen atoms per cubic centimeter. In other words, the corona, although it was highly tenuous at this distance, did reach all the way from the sun to the earth and beyond!

It was a startling idea: The earth in its orbit around the sun moves within the sun's hot corona. The corona is not a limited blanket enveloping the sun the way our atmosphere envelops the earth; on the contrary, the corona fills the whole solar system.

The Corona in Motion

It took a while for Chapman's statement to sink in. When it did, I recalled Biermann's description, during a visit to Chicago, of the corpuscular radiation that blows comet tails away from the sun. Now there were apparently two bodies of solar vapor to think about: the steady corona and the stream of particles flowing out from the sun at high speed. This, however, was impossible. In a magnetic field one stream of charged particles cannot pass freely through another, and it was known that solar-system space was filled with magnetic fields. Therefore the corona and the solar stream could not be separate entities. They must be one and the same. The corona, behaving like a static atmosphere near the sun, must become a high-velocity stream farther out in space. How could this come about?

I examined the mathematics of the barometric law in more detail and saw that, in the absence of a large inward pressure from outside the solar system, the high-temperature corona must flow away from the sun. To find the nature of this flow I then applied the

"SOLAR-WIND SOCK," the tail of a comet, always points away from the sun, blown back by the high-speed stream of hydrogen in space. Comet Mrkos (*opposite page*) was photographed in August, 1957, with a five-inch camera on Palomar Mountain. Irregularities in the comet tail are probably caused by turbulence in the solar wind.

descent corona outside it should be cooler. But about 15 years ago Martin Schwarzschild of Princeton University and Biermann independently presented a now accepted explanation of the paradox of the corona's high temperature. The corona is so tenuous that it takes very little heat to raise its temperature. Schwarzschild and Biermann suggested

that the churning motions of the gas at the surface of the sun generate low-frequency waves that provide enough energy to heat the corona to a million degrees. The action is somewhat akin to that of the boy scout who, although his body temperature is only 37 degrees C., produces enough heat to start a fire by rubbing two sticks of wood together

until they reach a temperature of several hundred degrees.

Our theoretical calculations cannot give us really accurate numbers for the speed and density of the solar wind, because this would require precise knowledge of the temperature and density of the corona near the sun—for which we have only rough estimates. But if we

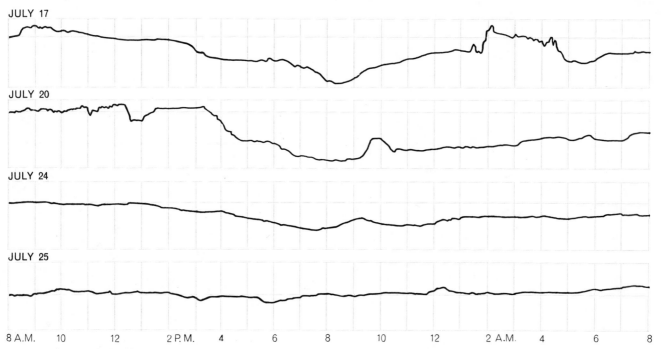

EARTH'S MAGNETIC FIELD fluctuates in response to changes in the solar wind as it blows past. These records of variations in strength of the horizontal component of the earth's field cover four full days during July, 1961. They were made in Honolulu.

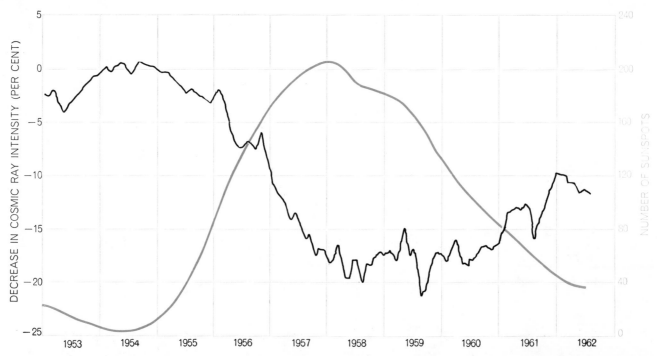

COSMIC RAY BOMBARDMENT of earth decreases as solar activity and wind increase during the 11-year "sunspot cycle." Black curve shows changes in cosmic ray intensity compared with maximum in 1954. Colored curve is a plot of solar activity as indicated by the number of sunspots. Spots are another manifestation of processes on the sun that cause fluctuations in solar wind.

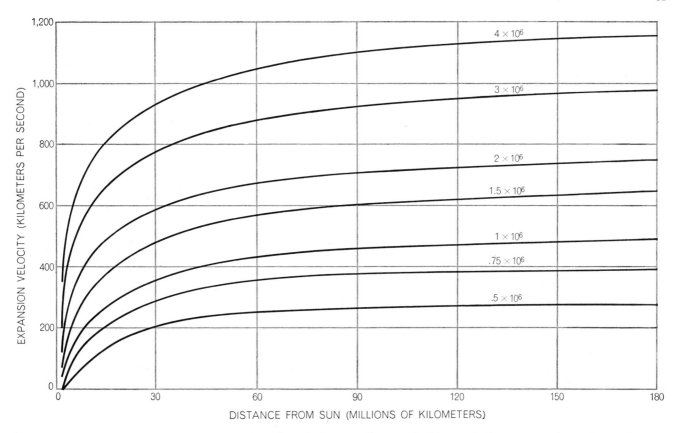

RATE OF EXPANSION of the solar wind into space depends in part on temperature of the corona. Temperatures are given on each curve in degrees absolute, ranging from 500,000 (*bottom*) to four million degrees. Orbit of earth is at 150 million kilometers.

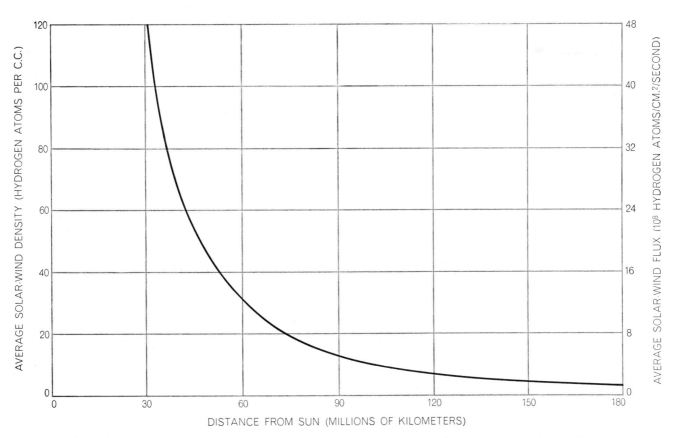

DENSITY AND FLUX of solar wind as function of distance from the sun are plotted from calculations. Scales are at left and right. The flux of the wind is defined as the number of hydrogen atoms passing through an area one centimeter square in one second.

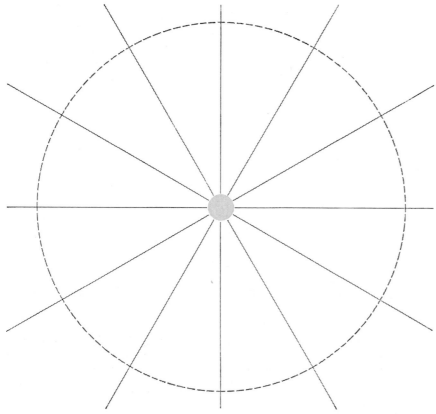

MAGNETIC LINES OF FORCE associated with the solar wind would appear as shown above if the sun did not rotate. The lines are in the equatorial plane of the sun. The broken circle marks the orbit of the earth, which is one astronomical unit from the sun.

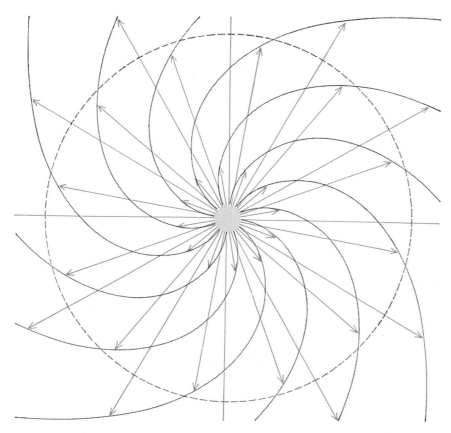

ACTUAL LINES OF FORCE are spirals due to solar rotation. They show how a compass needle would line up at any particular spot. The arrows are paths of solar-wind particles. Here the wind is assumed to be traveling through space at a steady 300 kilometers per second.

assume that the temperature at the base of the corona is a million degrees, we can draw the following approximate picture of the rise and progress of the solar wind. At the bottom of the corona the gas is almost stationary (in cosmic terms): it moves away from the sun's surface at the rate of only a few hundred meters per second. As it moves on it is replaced by more gas rising from below the photosphere. The coronal gas streaming slowly away is accelerated very gradually: it takes about five days and about a million kilometers of travel to get really under way. Thereafter it speeds up to hundreds of kilometers per second, and in four more days it has spanned the 93 million miles to the earth. The gas we see at the bottom of the corona on a Sunday will be passing us about Tuesday of the following week. Two weeks after this gas zooms by us it will pass Jupiter.

The Magnetism of the Wind

The solar wind carries a magnetic field along with it because the gas is ionized. (It remains ionized all the way out through the solar system, even though its temperature may drop to a low level; the gas is so tenuous that the separated protons and electrons have only a small probability of combining.) What is the nature of this magnetic field? Presumably its source is the general magnetic field of the sun. The corona cannot carry away the sun's concentrated local fields associated with sunspots and active regions, because these are strong enough to prevent the portions of the corona in their vicinity from streaming away at all. The sun's general field amounts to one or two gauss. (The earth's is about half a gauss.)

If the sun did not rotate (as it does once every 25 days), the solar wind would draw its general magnetic field straight out into space, so that the lines of force would stretch radially from the sun and a compass in solar-system space would always point straight toward or away from the sun. The sun's rotation, however, imposes on this radial field a circular field, with the result that the field carried by the solar wind takes a spiral form [see illustrations on this page].

The strength of a radial magnetic field, like that of gravity and light, weakens at a rate proportional to the square of the increase in distance from the source. It can be calculated, therefore, that at our distance from the sun the magnetic field carried by the solar wind should be down to about three or four hundred-thousandths of a gauss.

Evidence from Spacecraft

What have space probes shown about the solar wind? Many of the vehicles have carried equipment for recording charged particles encountered in space. In the first place, it can be said that they have unmistakably confirmed the existence of the wind. It was detected and measured by the Soviet vehicles *Lunik I* and *Lunik II* and by several U.S. vehicles, including the Venus craft *Mariner II* and the satellite *Explorer X*. They have shown that the wind blows continuously throughout the space they have traversed and that near the earth it is traveling at the expected velocity of about 400 kilometers per second. It blows straight out from the sun, sometimes steadily and sometimes in gusts. It tends to be turbulent and to move faster when the sun is active. The density of the wind has been hard to pin down. *Lunik I* and *Lunik II* indicated a flow rate of perhaps 100 million protons per square centimeter per second. *Explorer X* and *Mariner II* found that the wind's mean density near the earth lies in the range of one to 10 protons per cubic centimeter most of the time. This is in accord with the model of the corona that assumes that its temperature is close to a million degrees throughout the gas for a considerable distance from the sun.

In addition, the space vehicles' measurements of the magnetic field in interplanetary space bear out the theoretical picture of the solar wind. *Mariner II* and *Pioneer V* measured the field as being a few hundred-thousandths of a gauss, and *Mariner II* indicated that on the average the field had the expected spiral pattern. There were kinks and wiggles in the observed pattern, but this would merely confirm that the solar wind is sometimes gusty.

Fortified with all these confirmations of the nature of the solar wind and with some definite measurements, we can proceed to explore several interesting questions. For example, there is the matter of how much energy and mass the solar wind carries off into space. It can be calculated that it removes hydrogen from the sun at the rate of about a million tons per second. This is not a significant drain on the sun; in the estimated 15-billion-year lifetime of the sun it would amount to only a little more than a hundredth of 1 per cent of the

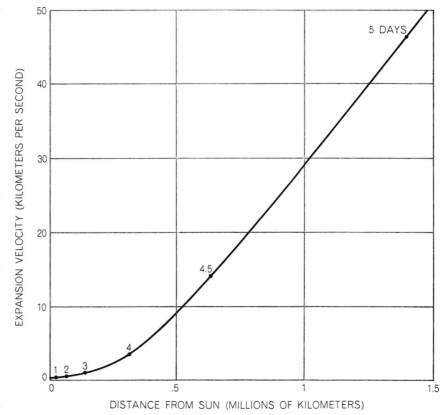

EXPANSION VELOCITY of solar corona near the sun increases rapidly after a relatively slow start. This is because the particles from the sun meet no resistance in space.

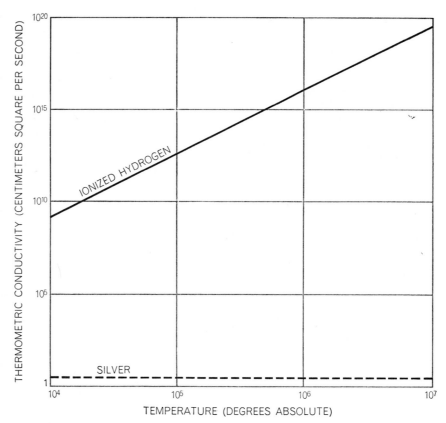

ABILITY TO CARRY HEAT of the solar wind is indicated by the thermometric conductivity of ionized hydrogen at a pressure typical of lower corona. Broken horizontal line is approximate value for solid silver, which could not exist at such temperatures.

solar mass. Similarly, the energy consumed in expanding the corona to the solar-wind velocity is only about a millionth of the total energy output of the sun. The wind's energy per unit of volume is so slight that an object in space is not warmed significantly by it.

How Far Does the Wind Blow?

There is also the question of how far the solar wind goes into space. This is considerably more interesting than the drain on the sun because it offers the possibility of using the solar wind as a probe into interstellar space.

The density of the wind must drop off in proportion to the square of the increase in distance from the sun. Eventually the wind must become so tenuous that it is stopped by the other thinly dispersed gases and weak magnetic fields in interstellar space. The general magnetic field of space in our galaxy is estimated to be no more than two hundred-thousandths of a gauss. If we take this maximum figure as the strength of resistance to the solar wind and use as our index of the wind's density the smallest value that has been measured near the earth (one atom per cubic centimeter), we can calculate that the solar wind ends at about 12 astronomical units (12 times the sun-earth distance) from the sun—that is, a little beyond Saturn. At the other extreme, if we take the smallest estimate of the resisting magnetic field (one two-hundred-thousandth of a gauss) and the highest measurement of the wind's density near the earth (10 atoms per cubic centimeter), then the wind goes out to 160 astronomical units —four times the distance of the farthest

planet, Pluto. These, then, seem to be the lower and upper limits: the solar wind apparently extends to at least 12 but not more than 160 astronomical units from the sun.

Two possibilities are at hand for exploring the outer limits of the wind. One is based on the fact that the hydrogen in interstellar space has been observed to emit faint ultraviolet radiation when it is excited. A recent analysis of such emission by Thomas N. L. Patterson, Francis S. Johnson and William B. Hanson of the Graduate Research Center of the Southwest in Dallas, Texas, suggests that the solar wind ends at perhaps 20 astronomical units from the sun.

The second possibility stems from the fact that the solar wind's magnetic field tends to sweep cosmic rays out of the solar system. During the years of high solar activity the intensity of cosmic rays coming to the earth is cut at least in half. We have calculated that a reduction of this size means that the solar wind extends well beyond Jupiter (five astronomical units from the sun). Simpson recently presented direct evidence that it probably goes out to at least 40 or 50 astronomical units. Analyzing the decline and recovery of cosmic ray intensity during the 11-year sunspot cycle, he found that the increase of the intensity of the higher-energy cosmic ray particles lags at least six months behind the drop in solar activity. The time lag apparently is a measure of the distance to the farthest extent of the solar wind. Just as it takes a certain time for a given ripple started in the middle of a pond to reach the edge of the pond, so will it take a certain time for an increase or decrease in the strength of the solar wind

to be communicated to the outer boundaries of the wind. Therefore there is a delay between a drop in the sun's activity, with the consequent weakening of the solar wind, and the arrival of the weakened wind at the limit of the space in which it acts as a barrier to the entry of cosmic rays into the solar system. Since Simpson finds the delay to be at least six months, and the wind travels at the rate of one astronomical unit in four days, simple computation shows that the distance to the borders of the solar wind is at least 40 to 50 astronomical units.

There is a great deal more to the observations than this one distance number. All in all, the fluctuations in cosmic ray intensity provide us with a natural probe for exploring the fields and other conditions in space out to the borders of the solar system and beyond, because the fluctuations bear the mark of the distant fields.

Do other stars have winds like the sun's? Very likely. The main requirement is that the star have a hot corona. Our sun's corona is generated by churning and convection of the gas beneath its photosphere. According to the theoretical picture of the interior of stars, subsurface convection is likely to occur in any ordinary hydrogen star with a surface temperature of less than 6,400 degrees. Most of the stars in our galaxy fall into this class, therefore stellar winds must be rather common.

The light from distant stars cannot tell us whether they have a corona or not. One really has to live in the midst of the wind to detect it. So most of our knowledge of stellar winds in general must come from studying the wind of the nearby star, the sun.

Comets in the News

10

INTRODUCTION

Exciting and significant cometary events are not limited to appearances of Halley's Comet or opportunities for space missions. Over the years, some of these events have been reported in *Scientific American's* "Science and the Citizen" department, and we present several of these items here.

In "Backward Comet" (October 1957), we find an explanation for the sunward tail, or anti-tail, of Comet Arend–Roland (Figure 1). The jets suggested by Fred L. Whipple may well exist in comets, but they are not responsible for anti-tails. The sunward tails are not truly sunward but merely appear so as a result of projection effects. When the earth is near the plane of the comet's orbit, which is the same as the plane of the solid material responsible for the anti-tail, the appearance resembles a spike, as shown in Figure 1. As the earth moves away from this plane, the appearance resembles a fan, as shown on page 45. Sunward appendages are not unique to Comet Arend–Roland; an anti-tail was observed, for example, in Comet Kohoutek in 1973.

"Trans-Neptunian Comet Belt" (August 1964) contains evidence for another storage area for comets. In more recent papers, Fred L. Whipple has revised the total mass in the possible comet belt down to not more than one earth mass

Figure 1. Comet Arend–Roland (1956h) on April 25, 1957, showing the "sunward spike." (Lick Observatory photograph.)

Figure 2. The signature of the Tunguska event. Trees were blown down up to 30 to 40 km from a central "ground zero" in a radial arrangement with the roots directed toward the center. This photograph was taken approximately 8 km from the center. (TASS from Sovfoto.)

for a belt at 50 AU. Even though the belt makes no appreciable contribution to the comets we observe, it may be a significant remnant from the processes responsible for the solar system's formation.

"Top View" (April 1976) shows a most unusual photograph of a meteor as observed from a satellite. Recall that most meteors are produced when cometary debris enters the earth's atmosphere and is heated and ablated by friction.

"The Comet Did It" (January 1961) is an early announcement that attributed a cometary origin to the Tunguska event (Figure 2). The details have evolved somewhat, but the cometary hypothesis is still the most promising explanation. Specifically, the latest suggestion is that a piece of Comet Encke about 100 meters across exploded in the atmosphere. Although the event was very destructive, the possibility of the earth being hit by a large chunk of a comet should not cause alarm or undue concern. First, only 1/100,000th of the earth's surface was affected by the blast. Second, chunks of the size responsible for the Tunguska event are expected to impact the earth only once in 2000 years on the average. Thus, the odds are very much against anyone presently alive being affected by such an event.

Backward Comet

Last April astronomers were treated to a rare sight: a new comet with a tail that points toward the sun instead of away from it, as most comets' tails do. Fred L. Whipple, director of the Smithsonian Institution's Astrophysical Observatory, now believes he can explain why this freak comet, called Arend-Roland, wears its tail backward.

The pressure of the sun's light is generally assumed to push a comet's tail away from its direction. Whipple suggests that Arend-Roland is probably a newly formed comet which has never before visited the neighborhood of the sun. Still unbaked by the sun's heat, it may have a feathery mantle of icy crystals. As it approaches the sun, heat may be penetrating deeply into this mantle

and vaporizing the crystals; jets of steam from the heated material would then shoot in the sun's direction, for the jets would be stronger than the pressure of the sun's light.

Trans-Neptunian Comet Belt

Slight perturbations in the orbit of Neptune, previously attributed to the gravitational attraction of neighboring Pluto, may in fact be caused by an invisible belt of comets that rings the sun just beyond the orbit of Neptune. This hypothesis was advanced recently by Fred L. Whipple, director of the Smithsonian Astrophysical Observatory,

in Proceedings of the National Academy of Sciences. As a corollary to his hypothesis, Whipple suggests that the mass of Pluto may be much smaller than has been supposed.

Current theories of the formation of the sun and the planets point to the existence of a large mass of small solid bodies at the outskirts of the solar system. The composition of these bodies is presumably similar to that of the outermost planets—Uranus, Neptune and Pluto—and these in turn appear to be comparable in composition to comets, which are believed to be icy conglomerates composed chiefly of frozen methane, ammonia and water. According to one theory the outermost planets are themselves accumulations of cometary material. Because of the low density of the outer parts of the primordial gas cloud, cometary material beyond the gravitational reach of Neptune would not coalesce into planets but would remain in a trans-Neptunian ring approximately in the plane of the planets.

Whipple estimates this belt of comets to have a total mass about 10 to 20 times that of the earth. If the diameter of the largest members of the belt does not exceed about 125 miles, roughly a thirtieth of the observed diameter of Pluto, they would have an apparent visual magnitude of 22 and be invisible to the largest telescopes. The belt would not contribute appreciably to the observable comets, which were probably formed inside the planetary region and thrown into their extremely elongated orbits after wandering into the gravitational domain of a massive planet such as Jupiter.

The gravitational effect on Neptune of such a belt of comets would account for the observed perturbations in Neptune's orbit more satisfactorily than the effect produced by a massive Pluto. Moreover, such a solution would avoid a long-standing dilemma regarding the mass of Pluto: if Pluto is responsible for the perturbations in Neptune's orbit, it would have to have a mass comparable to that of the earth, yet direct observations indicate a much smaller mass. By postulating the comet belt the mass of the more distant planet can now be considered to be as small as it appears to be.

Top View

From the ground a meteor streaking across the sky is seen as a celestial event. From above, however, it is seen as the kind of event it really is: a phenomenon of the earth's atmosphere. This was made vividly apparent to Captain John S. S. Kim of the Air Weather Service of the U.S. Air Force as he examined pictures made by one of the satellites of the Defense Meteorological Satellite Program. He describes his discovery in the magazine *Weatherwise*.

In the satellite photograph below, made on November 19, 1974, a meteor is seen as the streak of light in the lower left corner The bright spot at the upper right is the island of Oahu in the Pacific. At approximately the time when the satellite was overhead Air Force personnel stationed at Johnston Island saw the meteor from the ground as an unusually bright "falling star."

The Comet Did It

Investigators in the U.S.S.R. have come up with a new solution to an old geophysical mystery: What happened in the Tungus forest in Siberia on the night of June 30, 1908? The facts are that something exploded there, leveling many square miles of trees and knocking down people more than 30 miles away. At the same time the night sky glowed brightly over a large part of the Northern Hemisphere.

For years the explosion has been attributed to a great "Tungus meteorite."

Yet no one has found a crater in the area, or any meteoritic material. Last year the Soviet press published speculations, based on the supposed discovery of high radioactivity in the area by an expedition of amateurs, that the meteorite was an atomic-powered space ship. This publicity precipitated a new inquiry into the matter by the Committee on Meteorites of the Academy of Sciences of the U.S.S.R. The Committee now reports that there is no unusual radioactivity in the area and no meteoritic fragments.

According to Vasily G. Fesenkov, chairman of the Committee, the evidence indicates that the body which felled the trees was almost certainly a comet. In a summary prepared for *The New York Times*, Fesenkov said that the object approached from a direction opposite that of the earth's motion around the sun. Comets sometimes move in such a direction, but meteorites always travel in the direction of the planetary orbits. Moreover, the pattern of damage indicates an explosion hundreds of feet above the ground—a likely outcome for a body composed of dust and frozen gases, as comets are, but not for a meteorite of stone or iron. Finally, the extensive nightglow could have been caused by the dispersion of cometary material in the atmosphere. Taken altogether, the facts suggest the explosion of a comet several miles in diameter and weighing about a million tons.

A meteor (lower left) *is seen from a satellite*

BIBLIOGRAPHIES

Introduction

INTRODUCTION TO COMETS. John C. Brandt and Robert D. Chapman. Cambridge University Press, 1981.

1. Giotto's Portrait of Halley's Comet

THE NATURE OF COMETS. Nikolaus B. Richter. Methuen & Co Ltd., 1963.

THE COMPLETE PAINTINGS OF GIOTTO. Edi Baccheschi. Weidenfeld & Nicholson, 1969.

GIOTTO: THE ARENA CHAPEL FRESCOES. Edited by James H. Stubblebine. W. W. Norton & Co., Inc., 1969.

3. Comets

BETWEEN THE PLANETS. Fletcher G. Watson. The Blakiston Company, 1941.

A COMET MODEL. I. THE ACCELERATION OF COMET ENCKE. Fred L. Whipple in *The Astrophysical Journal*, Vol. 111, No. 2; pages 375–394; March, 1950.

5. The Tails of Comets

COMETS. Fred L. Whipple in *Scientific American*, Vol. 185, No. 1, pages 22–26; July, 1951.

6. The Nature of Comets

COMETS AND METEOR STREAMS. J. G. Porter. Wiley, 1952.

SURVEY OF THE UNIVERSE. Donald H. Menzel, Fred L. Whipple, and Gerard de Vaucouleurs. Prentice-Hall, 1971.

THE NATURE OF COMETS. Nikolaus Benjamin Richter (translated and edited by Arthur Beer; introduction and additional contributions by R. A. Lyttleton). Methuen, 1963.

THE MOON, METEORITICS, AND COMETS. Volume 4 of *The Solar System*, edited by B. M. Middlehurst and G. P. Kuiper. University of Chicago Press, 1963.

7. The Spin of Comets

THE PREDICTION OF ANOMALOUS TAILS OF COMETS. Zdenek Sekanina in *Sky and Telescope*, Vol. 47, No 6, pages 374–377; June, 1974.

BACKGROUND OF MODERN COMET THEORY. Fred L. Whipple in *Nature*, Vol. 263, No. 5572, pages 15–19; September 2, 1976.

8. The Origin and Evolution of the Solar System

ORIGIN OF THE SOLAR SYSTEM. Edited by Robert Jastrow and A. G. W. Cameron. Academic Press, 1963.

EVOLUTION OF THE PROTOPLANETARY CLOUD AND FORMATION OF THE EARTH AND PLANETS. V. S. Safronov. Israel Program for Scientific Translations, Jerusalem, 1972.

SYMPOSIUM ON THE ORIGIN OF THE SOLAR SYSTEM. Edited by Hubert Reeves. Edition du Centre National de la Recherche Scientifique, Paris, 1972.

NUMERICAL MODELS OF THE PRIMITIVE SOLAR NEBULA. A. G. W. Cameron and M. R. Pine in *Icarus*, Vol. 18, No. 3, pages 377–406; March, 1973.

EARLY CHEMICAL HISTORY OF THE SOLAR SYSTEM. Lawrence Grossman and John W. Larimer in *Reviews of Geophysics and Space Physics*, Vol. 12, No. 1, pages 71–101; February, 1974.

9. The Solar Wind

INTRODUCTION TO THE SOLAR WIND. John C. Brandt. Freeman, 1970.

DIRECT OBSERVATIONS OF SOLAR-WIND PARTICLES. A. J. Hundhausen. *Space Science Reviews*, vol. 8, pages 690–749; 1968.

DYNAMICAL PROPERTIES OF STELLAR CORONAS AND STELLAR WINDS: I. INTEGRATION OF THE MOMENTUM EQUATION; II. INTEGRATION OF THE HEAT-FLOW EQUATION. E. N. Parker. *Astrophysical Journal,* vol. 139, pages 72–122; January 1964.

DYNAMICS OF THE INTERPLANETARY GAS AND MAGNETIC FIELDS, E. N. Parker. *Astrophysical Journal,* vol. 128, pages 664–676; November 1958.

INTERPLANETARY DYNAMICAL PROCESSES. E. N. Parker. Interscience, 1963.

THE INTERPLANETARY MAGNETIC FIELD, SOLAR ORIGIN AND TERRESTRIAL EFFECTS. John M. Wilcox. *Space Science Reviews,* vol. 8, pages 258–328; 1968.

NOTES ON THE SOLAR CORONA AND THE TERRESTRIAL IONOSPHERE. Sydney Chapman. *Smithsonian Contributions to Astrophysics,* vol. 2, pages 1–12, 1957.

INDEX